The Plants of Caye Caulker

Jacob Rietsema
Dorothy Beveridge

Copyright 2009 Caye Caulker Branch, Belize Tourism Industry Association. All rights reserved. No portion of this book may be reproduced in any form or by any means, including electronic storage and retrieval systems, except by explicit, prior written permission of the publisher except for brief passages excerpted for review purposes.

Published for Caye Caulker Branch, Belize Tourism Industry Association by *Producciones de la Hamaca*, Caye Caulker, Belize

ISBN: 978-976-8142-22-1

The cover photo is of the cocoplum (***Chrysobalanus icaco***), which is the origin of the Spanish name for Caye Caulker, *Cayo Hicaco*. The top photo on p. 260 is by Daniel Atha. Photos on pages 18, 19, 25, and 27 are by James Beveridge. The bottom photo on page 234 is by Judy Lumb. The top photo on page viii is by Mo Miller. All other photographs are by Dorothy Beveridge.

Caye Caulker Branch,
Belize Tourism Industry Association
P.O. Box 43
Caye Caulker, BELIZE
Email: cayecaulkerbtia@gmail.com
Website: gocayecaulker.com

Producciones de la Hamaca is dedicated to:
—Celebration and documentation of Belize's
  rich, diverse cultural heritage,
  —Protection and sustainable use of Belize's
    remarkable natural resources, and
    —Inspired, creative expression of Belize's
      spiritual depth.

Acknowledgements

Without the help of a great number of people this book would never have been published. We wish to thank the villagers who let us collect plants from their gardens and shared with us their extensive knowledge of local plant names, as well as the medical and culinary uses: Petrona Joseph, Isela Marin, Lisa Novelo, Rico Novelo, Jr., Aurora Perez, Claudia Reyes, Lydia Vega, and Peter Young, Jr. Many other Caye Caulker citizens allowed us to collect specimens from their gardens, including Eulogia Aguilar, Efrain Aguilar, Elena August, Laura Badillo, Jessie Benner, Glenda Blease, Avril Cleland, Luciana Essenziale, Destino Fuller, Laura Hall, Emelda Heredia, Lionel Heredia (Chocolate), Sarah Heredia, Michael Joseph, Patricia Longsworth, Angelica Novelo, Efrain Novelo, Gertrude Novelo, Chris Roggema, Antonio Vega, Jr., Allison Wright, Alexander Young, and Shirley Young. We thank Paulina Cowo, Juana Cowo, and Sylvano Sho for giving us many Mopan Maya names for plants. We are grateful to Mo Miller for building the plant drying cabinet.

We thank the Board of the Caye Caulker Branch of the Belize Tourism Industry Association (CCBTIA) for their support and the use of the Resource Center. Board members over the five-year span of the project include Louis A. Aguilar, Adrian Allen, Dorothy Beveridge, David Heredia, Judy Lumb, Ernest Marin, Jr., Irene Miller, Javier Novelo, and others. All proceeds from the sale of the book will go to support the conservation work of CCBTIA.

The project was also endorsed by the national Belize Tourism Industry Association and the Forest and Marine Reserves Association of Caye Caulker, which allowed access to the Caye Caulker Forest Reserve.

We thank the Government of Belize, especially Chief Forest Officers Wilbur Sabido and Marcelo Windsor, for Research Collection Permit CD/60/3/06 (04) for the project, "An Inventory of Native and Introduced Plants of Caye Caulker" and export permits for sending samples for identification to the New York Botanical Garden.

We thank Hector Mai, the Administrator of the Belize National Herbarium, for his expertise and enthusiastic support of the project.

This work could not have been undertaken or completed without the support and help of Daniel Atha, Research Associate at the New York Botanical Garden. We thank him for his countless hours spent in plant determinations and for sharing his expertise in lectures and workshops during his visits to Caye Caulker.

We also acknowledge with appreciation the help of his assistant, Meryl Rubin, who patiently worked to make sense of our often poor notes to write herbarium labels for the samples we sent to the New York Botanical Garden. We must express our appreciation to the Staff of the New York Botanical Garden for their hospitality and access to the herbarium and library.

Our special word of thanks goes to Judy Lumb who brought this work to completion as a book. We are grateful for helpful reviews of the completed manuscript by Daniel Atha, Mike Amspoker, Ellen Armstrong, Philip Balderamos, James Beveridge, Isela Marin, Jan Meerman, Rico Novelo, Jr., Aurora Perez, and Lydia Vega.

We wish to thank Annelieke Schauer and Jeanne Overstreet for financial contributions.

Thanks to all.

Jacob Rietsema
Dorothy Beveridge

Foreword

In the summer of 2005 the Board of Directors of the Caye Caulker Branch of the Belize Tourist Industry Association (CCBTIA) was approached by one of its members, Dorothy Beveridge, with a proposal to prepare an inventory of the plants that grow on Caye Caulker. She had secured help from Dr. Jacob Rietsema, a retired botanist who every year spent several winter months on the island. Such an inventory of plants did not exist.

The Board noted that the original plant cover of Caye Caulker is rapidly disappearing due to increased development. Before more losses occur, knowledge of what grows on Caye Caulker should be documented. Such knowledge can help to preserve the natural environment for local citizens and tourists, and it will also help save part of our heritage since plants were, and still are, an integral part of the lives of the citizens, both as food and for medicinal purposes. It should stimulate a growing awareness of the importance of biodiversity and the need to preserve as much as possible.

Based on these considerations the Board agreed to support and sponsor this project and has continued to do so. Now, after five years of work, the project has been completed and resulted in this volume.

On behalf of the Board of Directors I am pleased to recommend this work to local citizens, interested tourists and botanists everywhere. The Board also expresses its appreciation to the authors who donated countless hours of work for the benefit of our community and to the villagers for their wisdom in choosing to preserve enough of a traditional "green" lifestyle, which provided the living laboratory for this book.

Louis A. Aguilar
CCBTIA Chairperson

Contents

Acknowledgements iii
Foreword v
Chapter I Introduction 1
    Caye Caulker 2
    Plant Structure 7
    Plant Identification 14
    How to Use this Book 15
Chapter II Underwater Plants 17
Chapter III Mangroves 21
Chapter IV Wetlands 29
Chapter V Littoral Forest 43
Chapter VI Roadsides and Open Areas 59
Chapter VII Gardens 113
Chapter VIII Trees 171
Chapter IX Palms 209
Chapter X Climbing Plants 217
Chapter XI Grasses and Sedges 235
Glossary 261
Inventory 269
References 278
Index 280
CCBTIA 291
Authors 291

Collecting plants in the Caye Caulker Forest Reserve for the CCBTIA Plant Inventory Project: (*above from left*) Jacob Rietsema, Ellen McRae, Lilliana Marin, Judy Lumb, Louis A. Aguilar, Dorothy Beveridge, Tony Vega, Sr., Allie Johnstone, Alex Young. (*photo by Mo Miller*)
(*below from left*) Louis A. Aguilar, Ellen Armstrong, Jacob Rietsema, Daniel Atha, Dorothy Beveridge, Mo Miller.

CHAPTER I Introduction

Without plants there cannot be life as we know it.

Plants make all life possible on earth. They harness and transfer energy from the sun. They produce oxygen and absorb carbon dioxide. They feed us and animals. They can protect us from sunlight, they can cool our houses, they protect us against bad weather, and they provide us with building materials. They have important medicinal properties. They enhance the quality of life by their beauty. They are indispensable for our well being.

In our efforts to preserve our natural environment we need an accurate description and inventory. This is basic to all future work. To this end the Caye Caulker Branch of the Belize Tourism Industry Association (CCBTIA) has undertaken to prepare an inventory of native and non-native plants found on Caye Caulker.

We began in 2005 and continued through 2009 collecting samples for identification. Plants were photographed and samples collected and catalogued. Samples were placed in a plant press and heated in a plant cabinet until dry. After getting proper permissions, the samples were shipped to the New York Botanical Garden for identification.

We are confident that we have collected or seen the great majority of plant species, but we are equally certain that new plants will be found and that plants once observed may disappear for some time and, possibly, reappear.

In this book we have given you an overview of what grows here. Please thank the villagers for their traditional lifestyle that has preserved our beautiful island, enjoy the beauty and importance of plants, and help protect our environment.

Caye Caulker

Geography

Caye Caulker is a reef shelf island situated in a shallow sea lagoon 1.5 kilometers west of the Belize Barrier Reef and 33 kilometers north-northeast of Belize City. The Caye is eight kilometers long, which is determined by two breaks in the Reef adjacent to the northern end (North Cut) and southern end (Caye Caulker Cut). The island is composed of a sand ridge surrounded by fringing mangroves which rest on a limestone base. An underwater cave system runs under the Caye. In the sand there is a lens of freshwater over the saltwater that allows for shallow wells.

Before the village was established, the Caye was surrounded by mangroves. A sand ridge runs north to south in the central area. Since the sand ridge is high enough to prevent it from frequent flooding, a littoral forest developed, consisting of woody shrubs and trees. Wetlands on the west (leeward) side of the sandy ridge include both permanent lagoons and seasonal lagoons present during the rainy seasons.

The island is divided into two parts by a channel referred to as the "Split". A low, narrow area of the Caye was dug deeper to allow boats to pass. This channel continues to be enlarged by currents and hurricanes.

Climate

Caye Caulker is sub-tropical, with temperatures ranging from 65 to 95 degrees F. There are four seasons, alternating wet and dry, but not of equal length. From October through January rain comes from the north when the cold fronts crossing North America reach Belize. The long dry season begins around February when the Trade Winds from the east bring warm, breezy days. The rainy season from June through early August is characterized by thunderstorms or squalls coming from the east. The next dry season is about six weeks in August and September, called the "Mauger Season", in which the wind drops and the temperature rises. While the Atlantic hurricane season is from June through November, it is in the latter part of the season when hurricanes have hit Belize.

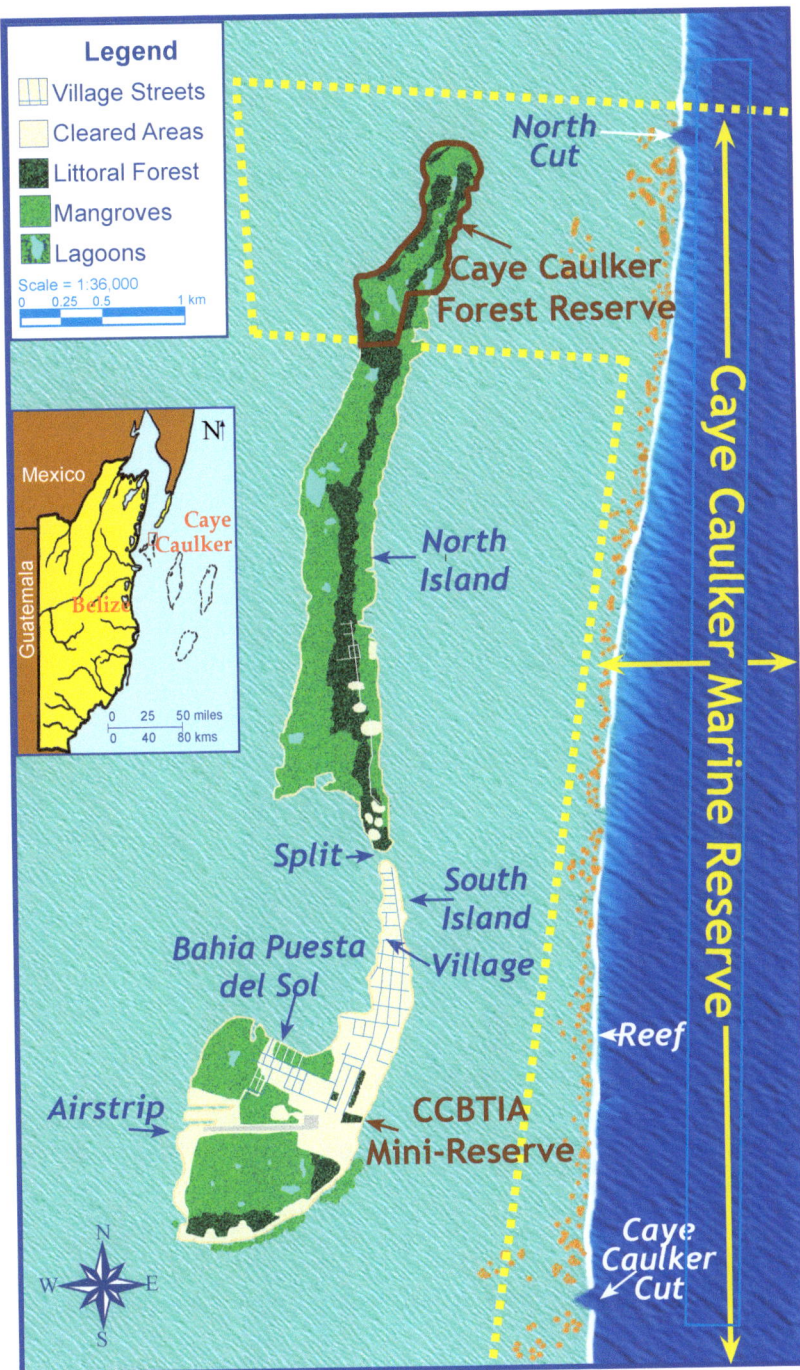

Figure 1. Map of Caye Caulker

History

The first permanent settlement on the island was founded in the mid-nineteenth century by Mestizo refugees from the Yucatan Peninsula. The village was established on the narrow high sand ridge in the middle of the island. The northern and southern high ridges were coconut plantations, called "cocals". The export of dried coconuts (copra), ship building, and fishing were the main industries up until the mid 20th century. Changing world economy, destruction of the coconut palms by hurricanes and disease hastened the demise of the copra industry. In the 1920s, the first lobster was exported and in 1960 the Northern Fisherman's Cooperative Society, Ltd., was established on Caye Caulker. The fledgling tourism industry began in the 1970s with local families renting rooms and cooking meals for the traveling "hippies". By the 21st century, tourism was the main industry on the Caye with continual development expanding into the wetlands. The village population is over 1800 which now represents all the ethnic groups of Belize, along with a number of people from other nations.

The Effect of Development on the Vegetation

Caye Caulker has a number of widely divergent types of environment, both natural and the result of development. The CCBTIA Mini-Reserve is a remnant of the original littoral forest. The seashores of the west and east side are different, because the east side is exposed to the predominant east wind and the west side is not. The mangroves of the east side are mostly gone, those on the west side less so. The coastal swamps of the southernmost area of the island are still only partly developed and remnants of the black mangrove forest are still present on the south side of the airstrip. The result of this wide divergence of growing conditions is a large variety of plants.

The native flora has evolved to flourish in sandy dry/wet soils with a high salt content in the air and soil. Natural conditions such as droughts, excessive rains, or hurricanes, as well as human development—dredging, clearing, filling, and building structures—alter the natural habitats resulting in changes in the flora of Caye Caulker. When plants disappear,

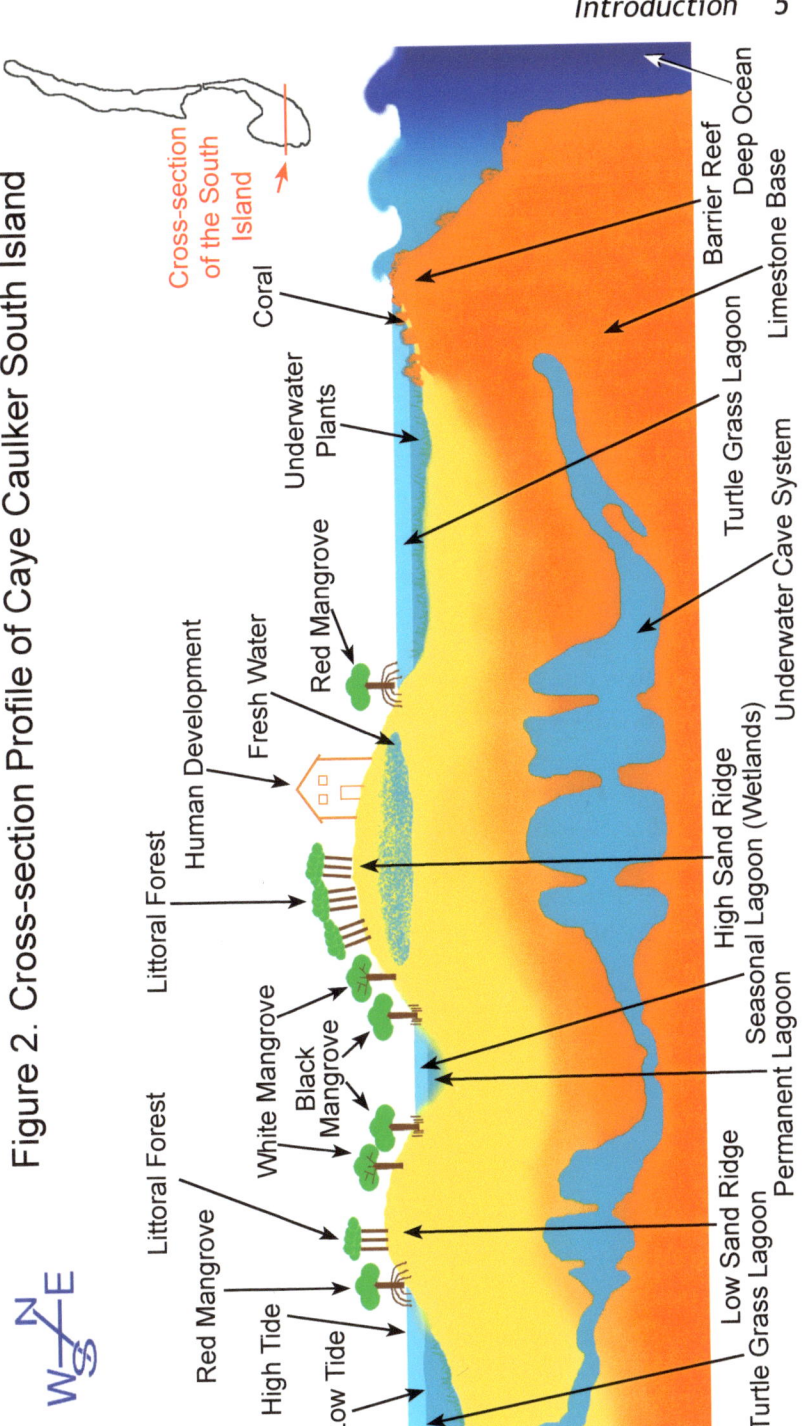

Figure 2. Cross-section Profile of Caye Caulker South Island

birds, lizards, and some insects also vanish. Habitat destruction leads to impoverishment of natural diversity.

Human development causes much of the native flora to disappear but also introduces foreign plants. This non-native flora consists of plants brought in for ornamental purposes, such as the hibiscus (*Hibiscus rosa-sinensis*) or for culinary or medicinal uses such as chaya (*Cnidoscolus chayamansa*) and aloe (*Aloe vera*). Many other plants came as fellow travelers with people coming here from the mainland or foreign countries. Seeds may be attached to clothing and luggage. Others are blown in by the wind, or come from the sea. Some became established and are now herbs, shrubs or trees accepted by the people as part of the normal flora.

In the village, on vacant lots, and along roadsides we find many herbaceous plants of foreign origin. In gardens many ornamental plants are found that often give colour to the streets and soften the hard lines of buildings. These habitats are not natural, but the consequence of human activities.

A fair number of plants are rare on Caye Caulker and only one or a few specimens have been found. Other plants are found in great number one year and seem almost extinct the next year. Building activity often destroys a location rich in plant diversity. Mowing of grass and broadleaf herbs in vacant lots is often fatal to herbaceous plants and low shrubs. Unfortunately these activities are expected to continue and are a serious threat to the diversity of the flora.

Almost all of the collected plants are from the southern island where development is the most intense. On the northern island we find littoral forest at the higher elevations, and coastal swamps and mangroves in the low lying areas. The flora on the northern island resembles the original plant cover.

Not so very long ago many islanders depended on medicine from herbs and a number of herbalists practiced, helping people with medicines from local plants. Recently modern medicine with pills for every disease has decreased the dependence on herbs, particularly among younger people. There are, however, a few herbalists who still practice the traditional medicine.

Reserves

The ecology of the island is very fragile and is easily changed by nature and man. To help protect remnants of the original flora and fauna, reserves have been established on Caye Caulker. The Caye Caulker Forest Reserve is 100 acres of littoral and mangrove forests on the northern tip of Caye Caulker. It was established in 1998 under the Forest Act and is co-managed by the Forest Department and the Forest and Marine Reserve Association of Caye Caulker (FAMRACC), a local co-management association. The seven-mile long Caye Caulker Marine Reserve includes the Belize Barrier Reef and adjacent grass flats off shore from Caye Caulker. It was established in 1998 under the Fisheries Act and is co-managed by the Fisheries Department and FAMRACC.

The Caye Caulker Mini-Reserve is 1.9 acres of littoral forest and coastal mangroves. It was established in 1996 and is managed by the Caye Caulker Branch of the Belize Tourism Industry Association (CCBTIA).

Plant Structure

Parts of Plants

A plant consists of **roots** which are usually in the ground and the **stem**, which is usually above the ground. The stem can be soft and green as in **grasses** and **herbs**; or the stem can become woody as in **shrubs** and **trees**. The stems of herbs and the woody twigs of shrubs and trees bear the **leaves** and the **flowers**. Plants are identified and classified by their structure, which is the subject of much study by botanists. We present a mere summary to explain terms that are used in the plant descriptions.

Bark

The outside of the woody stem is the bark which can be gray, black, brown, or splotchy, can be smooth, rough, or fissured, or it can peel. Out of cracks in the bark from some plants a resin flows or oozes, which can occur naturally or as a result of a wound.

Leaves

To identify a plant one begins with the leaves, the first thing that you see. The place where leaves are attached to the stem is the **node**. The stem attaching the leaf to a branch is the **petiole**. The petiole can be almost absent or very long. The angle of the petiole with the stem is the **axil**. There are three types of arrangements of **leaves** (*Figure 3*) on stems.

Single leaves may be **alternate**, which means that they are attached at different nodes on the stem, either across in a flat configuration, or spirally around the stem.

Two leaves may be attached directly **opposite** at the same node, or

Three or more leaves may be **whorled**, which means they are attached around the same node.

A vertical row of leaves (or flowers) is a **rank**, so plants with two rows of leaves on either side of a stem are **two-ranked** and those with three vertical rows are **three-ranked**.

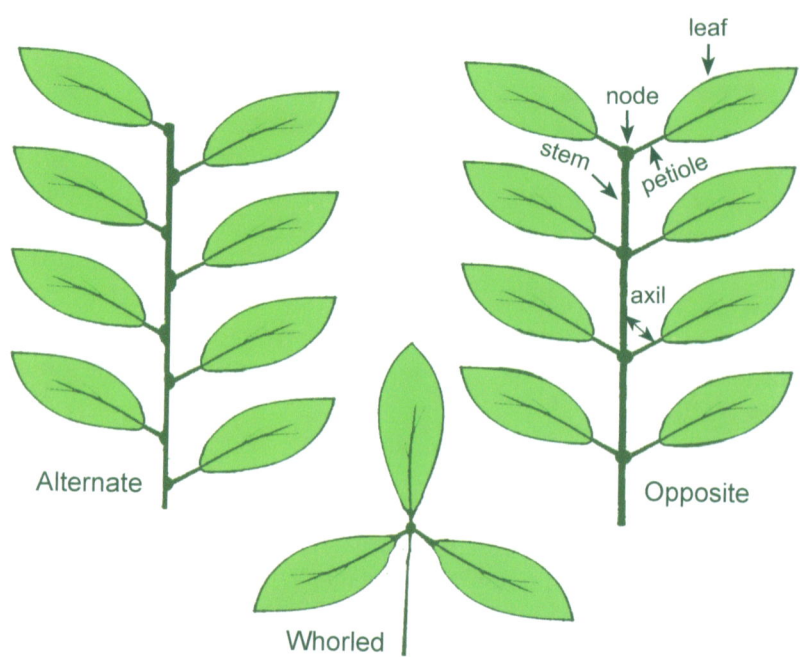

Figure 3. Arrangements of simple leaves.

Introduction

Compound leaves (*Figure 4*) are divided into three or more **leaflets** in various branching configurations. When three leaflets radiate from a single point they are **trifoliate**. When more than three leaflets radiate from the axis, the configuration is called **palmate**. When the leaflets are in two rows along the stem, the configuration is **pinnate**, which can have either an **odd** or **even** number of leaflets, always opposite. Leaves can be pinnately divided twice (**bipinnate**) or three times (**tripinnate**).

Leaves are of various shapes (*Figure 5*), which are important characteristics of a plant to assist with identification. Leaves can be **linear, lanceolate, ovate, elliptic, orbicular, oblanceolate, obovate, spatulate, or oblong.**

Tips (*Figure 6*) of leaves can be tapered with a sharp point (**acuminate**), rounded with a point (**acute**), round (**obtuse**), indented (**retuse**), or rounded with a little point at the tip (**cuspidate**).

Bases (*Figure 7*) can be tapered (**attenuate**), wedge-shaped (**cuneate**), round (**obtuse**), or heart-shaped (**cordate**).

The leaf edge (*Figure 8*) can be smooth (**entire**), indented (**lobed**), toothed with rounded teeth (**crenate**), sharp teeth (**serrate**), or fine sharp teeth (**serrulate**), or wavy up and down (**undulate**).

Flowers

Flowers are made up of leaves that grow into a particular shape, colour, or form for the purpose of seed production. The colour of flowers and the fragrances they produce attract pollinating agents, such as bats, birds, butterflies, and bees. For example, red flowers attract hummingbirds. Bats are active at night, so plants that attract bats bloom at night and mostly have white flowers.

Flower structure (*Figure 9*) is complex and highly variable, but the basic structure includes the female parts (**pistil** composed of **stigmas, style,** and **ovary**), the male parts (**stamens** composed of **anthers** and **filaments**), the **corolla** (composed of the **petals**) and the **calyx** (composed of the **sepals**). **Inflorescence** is the appearance and arrangement of flowers, which is easier to observe and quite helpful for identification of plants (*Figure 10*).

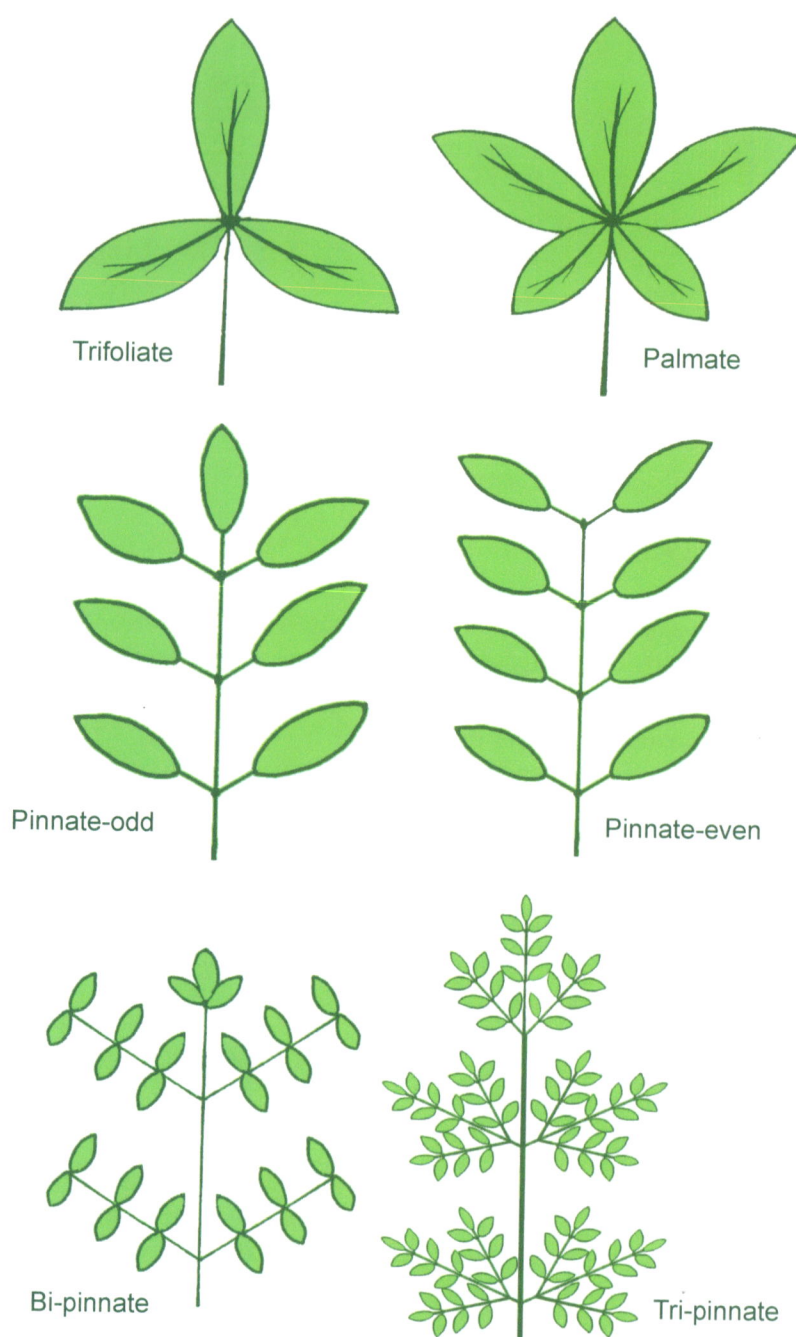

Figure 4. Arrangements of compound leaves.

Introduction 11

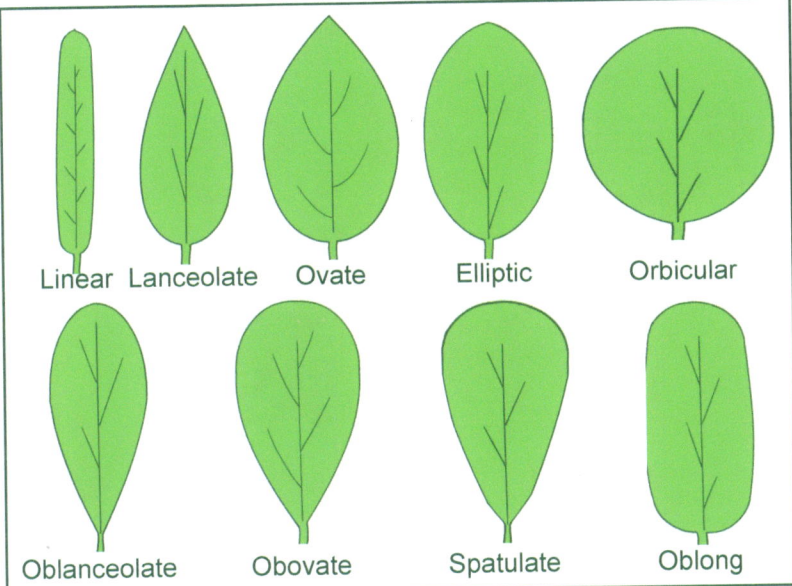
Figure 5. Shapes of leaves.

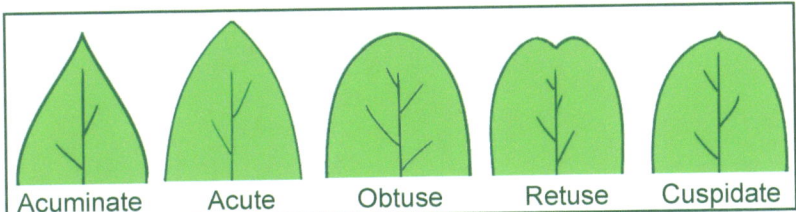
Figure 6. Tips of leaves.

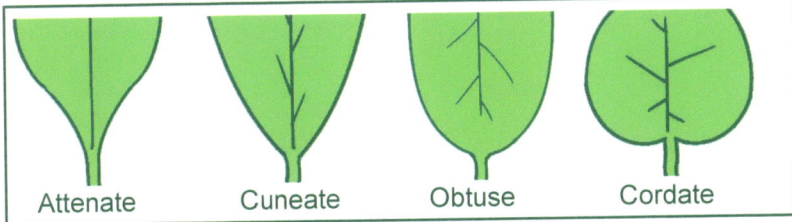
Figure 7. Bases of leaves.

Figure 8. Edges of leaves.

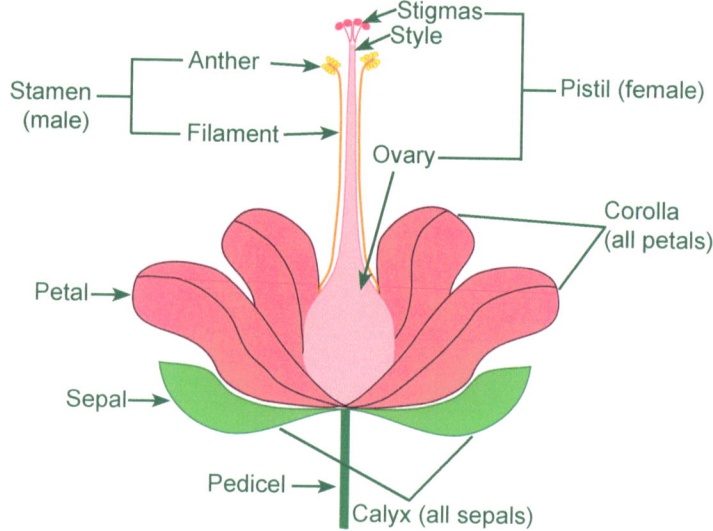

Figure 9. Flower Structure

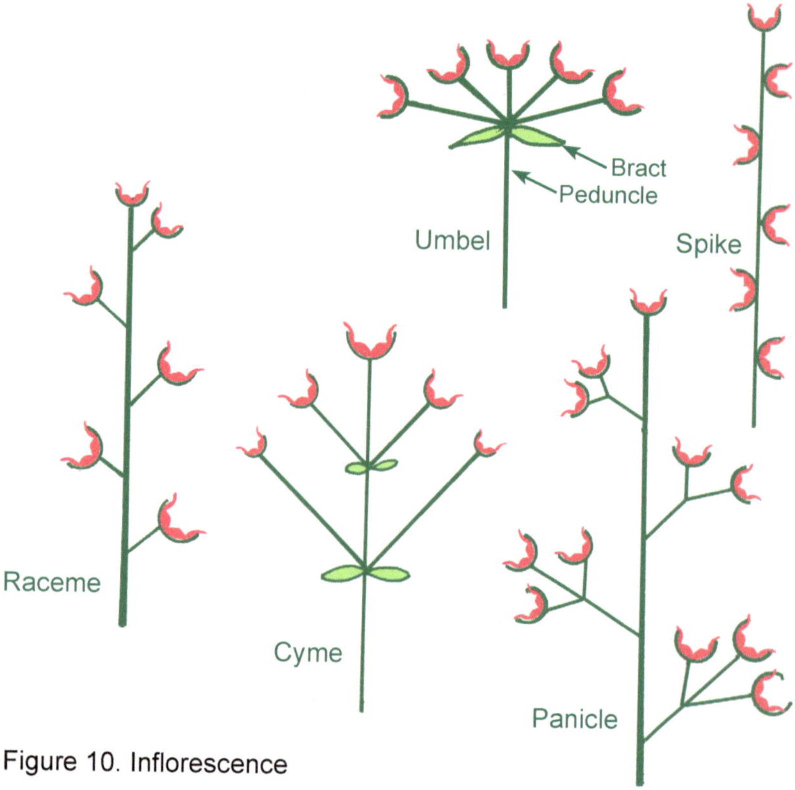

Figure 10. Inflorescence

Introduction 13

Bracts are leaves that envelope flowers or an inflorescence. A single flower is attached by a **pedicel**, while a flower cluster is attached by a **peduncle**, both of which are absent in some plants. The simplest inflorescence is a single flower at the end of stem but mostly the inflorescence is composed of many flowers in various arrangements. A flower arising from the axil of a leaf, where the leaf meets the stem, is called "**axillary**." When multiple flowers occur at the end of the stem with all pedicels arising from the same point, it is called an "**umbel**." An inflorescence that occurs along an unbranched stem without a pedicel is a **spike**. If the stem is branched and the flowers are attached directly to the branches without a pedicel, the branches are **spikelets**. Multiple stems attached to an unbranched stem by a pedicel form a **raceme**. A **panicle** has a branched stem with branches that are also branched.

The order in which flowers bloom along a stem is important. A **determinate** or **cymose** inflorescence is when the central flower at the end of the stem blooms first, followed by the neighboring flowers going down the stem. In cymose plants (**cyme**), the oldest flower is at the end of the stem, which will not grow further. In plants with **indeterminate** or **racemose** florescence, the bottom flowers are first, followed by those going up the stem, so the stem keeps growing.

Flowers of some families are even more complex—what looks like one flower is actually a flower head (**corymb**) composed of many flowers. For example, in the *Asteraceae*, the family to which rabbits paw (*Sphagneticola trilobata*) belongs, the flowers at the edge are the **ray flowers** and at the center are the **disc flowers**. The disc flowers are usually yellow, but they are very small and close together. The ray flowers actually determine the colour that can be seen.

Fruits and Seeds

The **ovary** of a flower develops into the fruit. The **ovules** inside the ovary become the **seeds**. Fruits can take many sizes and shapes from a soft berry, to a stone fruit with a hard pit in the center, to a capsule which is dry and can split open. Seeds are found in the fruit. Some seeds grow together with the fruit to form the **nut**. Bananas that we eat are large berries, but the seeds have been lost in cultivation.

14 Plants of Caye Caulker

Coconuts *(Cocos nucifera)* are actually nuts, that is, both fruit and seed. The outer green skin, the fibers just inside, and the outer layer of the nut inside are the components of the fruit. The seed is the inner layer of the hard shell and the meat that we eat. Here we also find the embryo. The three black spots on the coconuts are places through which the embryo will grow upon germination.

Plant Identification

The plants described in this book have been photographed and collected by the authors. All photographs have been made of plants as they grow on Caye Caulker. With few exceptions the collected samples have been shipped to the herbarium of the New York Botanical Garden (NYBG) where they were identified, mounted and returned to Belize. The mounted samples were added to the National Herbarium collection in Belmopan. Final identification was made by Daniel Atha, Research Associate, NYBG. In most cases duplicate samples were added to the NYBG Herbarium collection.

The scientific (Latin) names are very important because no two plants bear the same recognized scientific name. When two or more plants are identical and usually can be cross-pollinated, they are said to belong to the same species. When two or more species are similar, but usually cannot be cross-pollinated, they are considered to belong to the same genus. Scientific names are italicized; the capitalized genus name is first, followed by the species name which is all in lower case. The author of the scientific name is indicated by initials or abbreviated names behind the species name. For example, *Cordia sebastena* L. is the scientific name for zericote. The genus is *Cordia*, the species is *sebastena* and L. stands for Linnaeus (Linne) who gave the plant its current legitimate name.

Related genera (the plural of "genus") are grouped into families, the names of which end in "-aceae". For example, zericote (*Cordia sebastena*) is in the family *BORAGINACEAE*. In the descriptions of the plants, the family names are found in the upper right, all in capital letters and italicized.

Introduction   15

The *FABACEAE* family is divided into three subfamilies, which end in "-oideae". The subfamilies are on the same line, but not all in capital letters. For example, for the beach pea (*Crotalaria verrucosa*) the "family: subfamily" is indicated as "*FABACEAE: Papillionoideae*".

For some plants the species name could not be determined with accuracy and the genus name is given, along with "sp." to indicate a species of that genus. For example, the ornamental fig for which the species could not be determined, is listed as *Ficus* sp. The nomenclature used is that of the *Checklist of Plants of Belize* by Balick *et al*, 2000.

Plant common names (Creole, English, Spanish, Mayan, or Garifuna) used on Caye Caulker were obtained from local inhabitants and are indicated in **Bold** type. Other common names are not in bold. A plant may have different common names, even in the same locality, or different plants may have the same common names, so common names can be confusing.

How to Use this Book

The plant descriptions in the following chapters are presented to assist local citizens and visitors with plant identification and to increase the appreciation of the local flora.

The separation of plants into chapters is quite arbitrary, so it may be necessary to look in more than one chapter to find a particular plant. Most plants are described in chapters based upon where they are likely to be found, their habitat: **Underwater Plants, Mangroves, Wetlands, Littoral Forest, Roadsides** and **Open Spaces**, and **Gardens**. The chapter on Gardens is organized according to the colour of the flowers. Within the chapters and sub-sections, individual plant descriptions are in alphabetical order by genus name.

The chapter on **Trees** includes the other trees that are neither palms, mangroves, nor native trees found in the littoral forest. Most are found in gardens.

Some plants are found in all habitats, not exclusively in any of the habitats listed above, so they are included in chapters based upon their appearance which is easily recognized. **Palms** can be recognized because they usually have unbranched stems ending in clusters of leaves, as seen in the coconut palm. **Climbing Plants** can be recognized because they climb on other plants or fences. **Grasses and Sedges** are low plants with bunches of long narrow pointed leaves.

Some plants have been collected or seen that at this time could not be considered permanent residents of Caye Caulker until they occur in more than one location and more than one season. Some plants were found for which a photograph or description was not available. These plants are included in the Inventory, but are not described.

The plant descriptions contain a minimum of technical terms, all of which are included in the Glossary. Sizes given are the usual maximum that a plant, tree, stem, leaf, flower or other plant part might attain when mature.

CHAPTER II
Underwater Plants

In the sea around Caye Caulker grow two species of plants called "sea grasses", although neither is a true grass. Both are flowering plants. The leaf blades are easily seen and provide food for marine animals, thus functioning as primary producers in a food web. The flowers are tiny and very rarely seen. There are several other species but only these two are abundant—turtle grass and manatee grass—both growing on the ocean floor. They are important for fish and other sea animals as breeding grounds. The grasses hold the sand in the lagoon, which also protects coral from damage due to siltation. Pieces of leaves and sometimes whole plants are washed ashore.

18   Plants of Caye Caulker

Syringodium filiforme Kütz.

CYMODOCEACEAE

Syn. *Cymodocea filiformis* (Kütz.) Correll

manatee grass, eel grass

Native Range: Atlantic and Old World tropical and subtropical coastal sea floor

Manatee grass is not a grass, but an herb with stems that run under the sea floor (rhizomes). The thin leaves (50 cm long) are long and cylindrical. Pieces are found washed up along the beach. Tiny flowers appear from January to June, but are seldom seen.

The rhizomes help prevent erosion of the sand on the sea floor. West Indian manatees (*Trichechus manatus*) graze on the plants, pulling them up by the roots.

Underwater Plants 19

HYDROCHARITACEAE

Thalassia testudinum **K.D. Koenig**

turtle grass

Native Range: Sea floor of Atlantic coast of Florida and Caribbean, South America

Turtle grass is not a grass, but an herb with long rhizomes. In contrast to manatee grass, the leaves (10-60 cm by 1 cm) are flat. The edges of the leaves are either smooth or minutely serrated. Flowers occur from January to June, but are rarely seen.

Green turtles (*Chelona mydas*) eat the leaves of turtle grass, which accounts for the name.

(*above*) Daniel Atha explains to the Caye Caulker environmental club the importance of studying plants and how they are collected.
(*below*) Annabella Requena puts a plant into a press to be dried and stored in an herbarium for future study.

CHAPTER III
Mangroves

Mangroves are critical for the protection of the Caye against hurricanes and heavy seas, preventing erosion of the higher land. Mangroves are unrelated species that are characterized by their mechanisms to deal with salt, either by filtering it out at the roots, or excreting it through the leaves or bark.

Mangroves are examples of plants that may grow as trees or shrubs. There are three mangrove species—red, white and black—that tolerate salty or brackish water and also grow on dry land.

Outermost, growing in the water, is the red mangrove characterized by prop roots. These densely tangled mangrove roots serve as a fish nursery, a safe haven for shrimp larvae, small fish, and shell fish.

White mangroves grow on land that is frequently flooded, such as, in the wetlands south of the Airstrip.

Black mangroves grow around permanent lagoons that stay wet most of the year.

Buttonwood, also called "grey mangrove", is usually associated with the other mangroves, but it grows on drier soil and is not tolerant of frequent flooding, so it is not considered a true mangrove species.

Two epiphytes grow in the mangroves, the cowhorn orchid, and a bromeliad. They are rarely seen because of the impenetrable mix of roots and branches that make a mangrove forest almost inaccessible.

22 Plants of Caye Caulker

VERBENACEAE

Avicennia germinans (L.) L.

black mangrove, mangle

Native Range: East and west coasts of tropical and subtropical America

Black mangrove is a large shrub or a small tree that is heavily and irregularly branched. The bark is dark on older trees. The leaves (4–12 cm long by 4 cm wide) may be oblong, elliptic, or lanceolate. The tips may be acute or blunt. The pubescent leaves are light green and grayish underneath. The leaves are opposite and attached by a short (3-15 mm) petiole. The small white flowers are tubular, usually in panicles. The fruits are green pods. The numerous breathing roots (pneumatophores) emerge from standing water and are usually taller than those of the white mangrove.

In 2000 Hurricane Keith heavily damaged the black mangrove forest. Illegal and legal, but inadvisable, clearing for development are both major threats.

COMBRETACEAE

Conocarpus erecta L.

buttonwood, grey mangrove

Native Range: Tropical and subtropical America and west coast of Africa

Buttonwood is a large (5 m) shrub. The leaves are lanceolate or elliptical (8 cm long), pointed at both ends and smooth. They are alternate and attached by short (3-10 mm) petioles. At the base of the leaves is a pair of salt glands. The green flowers occur in small (1.5 cm across) nearly round heads (buttons) that are in small clusters. The fruits have a brownish tint when ripe. The pneumatophores are short and stubby and often not present.

A variety with grey or silver leaves that are densely pubescent also grows on the caye and is sometimes used as an ornamental.

The wood is used for charcoal. The bark contains tannin. When Olive-throated Parakeets *(Aratinga nana)* come from the mainland they feed on the buttonwood fruit.

24 Plants of Caye Caulker

COMBRETACEAE
Laguncularia racemosa (L.) C.F. Gaertn.
white mangrove

Native Range: East and west coasts of tropical and subtropical America, west coast of Africa

White mangrove is a shrub or tree that grows up to ten meters on Caye Caulker and taller on the mainland. The leaves are elliptical (4-10 cm long by 2.5-5 cm wide), round at both ends, smooth along the edge, and opposite. There are two glands on the petiole near the origin of the blade (*photo insert*). The fragrant, whitish flowers (5 mm long) are in long panicles (3-10 cm long) at the ends of twigs. The fruit is flesh around a stony seed, which sometimes germinates while still attached to the plant. The pneumatophores are shorter and more stubby than those of the black mangrove. The white mangrove has no aerial or prop roots which distinguishes it from the red mangrove.

The leaves and bark are rich in tannin. Tall white mangroves from the mainland are used for masts on sailboats.

ORCHIDACEAE

***Myrmecophila* sp.**

Syn. *Schomburgkia* sp.

cowhorn orchid

Native Range: Tropical Central America

The cowhorn orchid is a large epiphytic orchid found in the mangrove forest. The leaves grow out of large yellow, hollow pseudo bulbs, which are usually associated with ants. Large yellow-orange flowers are produced on long stalks in May.

Collected plants are sometimes found in gardens attached to tree trunks. It is likely that this is *Myrmecophila tibicinis* (Bateman) Rolfe, but other species have been suggested.

RHIZOPHORACEAE
Rhizophora mangle L.
red mangrove, mangle

Native Range: Tropical and sub-tropical coasts of America and west coast of Africa

Red mangrove is a shrub that can grow up to ten meters tall, but on Caye Caulker most are much shorter. The leaves are opposite, elliptic (12 cm by 2.5–5 cm), and have pointed tips and bases. The leaves are attached to the branch by a pale yellow petiole (1.5–2 cm long). The inflorescence is a branched stalk (4–7 cm long) with two to four small yellow flowers (0.75 cm across) found in a leaf axil. The fruits germinate on the tree, producing roots as long as 30 centimeters before falling into the sea. The prop roots grow off the main stem, while aerial roots grow down from branches, both of which form the stilts that are characteristic of the red mangrove.

The bark and leaves are rich in tannin. The curved prop roots were used for the ribs of the traditional sailing fishing smacks.

Mangroves 27

BROMELIACEAE
Tillandsia streptophylla **Scheidw. ex E. Morren**

air plant

Native Range: Central America, Mexico, Caribbean

Air plant is a bromeliad with many grey triangular leaves that curl when dry. The sheaths (basal part of the leaves) grow from a round dense pseudo-bulb. Inflorescences with red stems grow on the flower stems. The purple flowers have papery bracts for wind dispersal.

This is the only *Bromelia* found in the mangroves on Caye Caulker. Because it is located deep in the mangroves it is not easy to see.

(*above*) Daniel Atha explains to Louis A. Aguilar and Dorothy Beveridge that it is important to spread the plants flat before putting them in the press.

(*below*) Isela Marin tells Daniel Atha and Jacob Rietsema about the uses of the calaloo (*Amaranthus dubius*).

CHAPTER IV
Wetlands

The wetlands habitat is found in the low lying areas of the Caye: the tidal water surrounding both the northern and southern islands, seasonal lagoons located on the west (leeward) side that flood during the rainy season, and the permanent lagoons found in interior areas.

Some wetlands have recently been subdivided into building lots on both the northern and southern islands. The Bahia Puesta del Sol development was once a swampy black mangrove forest. The surface water of such land is always brackish. Even when developed, the water table remains very close to the surface and not all plants will grow on such soil. With this conversion to building lots, plants such as saltwort and cattail disappear.

SCROPHULARIACEAE
Bacopa monnieri (L.) Wettst.

water hyssop, bramhi

Native Range: China, now widespread in subtropics and tropics

Water hyssop is a low mat-forming herb not more than five centimeters high. The succulent leaves (2 cm long) are oblong. The light blue flowers (1 cm across) have five sepals, five petals, and are barely two-lipped.

Water hyssop grows in abundance on flat, low-lying areas. It is an important medicinal herb, especially in India.

Wetlands 31

BATACEAE

Batis maritima L.

pickle weed, saltwort

Native Range: West coast of Neotropics, now also in Hawaii

Pickle weed is a low-growing, salt-loving plant that forms mats. The fleshy, yellow-green leaves are linear-oblanceolate, and arranged opposite. The inflorescence consists of very small white flowers on short spikes rising from the base of the leaves. Male and female flowers are on different plants.

This edible plant grows on coastal wetlands in sunny areas.

32   Plants of Caye Caulker

AMARANTHACEAE
Blutaparon vermiculare (L.) Mears.
saltweed, silverweed
Native Range: Central America, Mexico

Saltweed is a low, flat, fleshy, perennial herb (25 cm tall) with many branches. The leaves (5 cm long) are oblong to linear, and attached without petioles to reddish-pink branches. The small, white flowers bloom year round on very short stems, in globular or elongated short spikes.

Saltweed grows in coastal wetlands in open sunny areas. It is edible when cooked like spinach.

ASTERACEAE

Borrichia arborescens **(L.) DC.**

seaside oxeye, seaside daisy

Native Range: Southern Florida, Caribbean, Central America

Seaside oxeye is an evergreen shrub (1 m tall by 2-3 m wide) with a silvery appearance. The slightly fleshy grey-green leaves (8 cm long) are oblanceolate-spatulate and grow opposite directly on the branch. Each three-centimeter flower head is on a single stem (3-5 cm long). The ray flowers on the outside are pale yellow, while the disk flowers inside are orange-yellow. The black fruits are achenes (3-4 mm long).

It is a slow-growing plant that grows in coastal wetlands in open sunny areas.

34 Plants of Caye Caulker

BRASSICACEAE
Cakile lanceolata **(Willd.) O.E. Schulz**

sea rocket

Native Range: North America, Mexico, Central America

Sea rocket is an upright or spreading, annual or perennial herb (80 cm tall). The succulent leaves, ovate-obovate-spatulate, and deeply lobed. They are attached to the branch with a distinct petiole. The inflorescence is a raceme. The white or slightly lavender flower has four spreading petals (0.5 cm across). The fruit (20- 25 mm long by 5 mm wide) is divided into two segments, the lower part being shorter than the upper part.

Sea rocket grows close to the sea in open areas.

ASTERACEAE

Eclipta prostrata (L.) L.

false daisy

Native Range: Probably Asia, now worldwide.

False daisy is an annual herb that grows to 50 cm tall. The hairy leaves (10 cm long) may be lanceolate, elliptic or oblong, pointed at both ends, with or without petioles, smooth or somewhat serrated, and grow in an alternate configuration. The white, roundish flower head (1 cm wide) can be either at the end of the flower stem (7 cm long) or where the flower stem meets the branch. The whitish ray flowers on the outside have only the female parts, while the yellow disc flowers inside have both male and female parts.

False daisy grows in wet, poorly drained areas.

36 Plants of Caye Caulker

BORAGINACEAE

Heliotropium curassavicum L.

seaside heliotrope

Native Range: Tropical and subtropical Americas, now India and Australia

Seaside heliotrope is a short-lived perennial herb (50 cm). The blue-green fleshy leaves (3.5 cm long) are linear to obovate or spatulate with short petioles and smooth edges. The flowers (5 mm long) are white with a yellow centre, but turn blueish. They are in unbranched coiled cymes. The fruit separates into four nutlets.

Seaside heliotrope grows on saline or alkaline soils in open areas, but is uncommon on Caye Caulker.

APOCYNACEAE

Rhabdadenia biflora (Jacq.) Müll. Arg.

Christmas flower, mangrove rubber vine

Native Range: Florida, Caribbean, Central America, northern South America

The Christmas flower is found climbing or spreading on the ground. The leaves (10 cm long by 2-3 cm wide) are obovate, lanceolate, or oblong with a sharp tip. Leaves are attached to the stem by petioles (1-2 cm long). The inflorescence is a small cluster of three to five flowers that are laterally attached to the stems. The flower looks like a white morning glory with a two-centimeter tube and a three-centimeter-wide conical throat with five lobes. The long narrow fruit pod (10 cm long) has one-centimeter-long seeds with tufts of long hairs.

Christmas flower grows in flat, low-lying, sunny areas.

38   *Plants of Caye Caulker*

CHENOPODIACEAE
Salicornia bigelovii **Torr.**
glasswort
Native Range: East and west coasts of North America, Caribbean, Mexico

Glasswort is a low (30 cm) branched erect annual herb that is green, but turns red in the dry season. The succulent leaves are reduced to fleshy bracts. The flowers are extremely small on cylindrical spikes (10 cm long by 6 mm wide), which are almost concealed by the bracts. Of the three flowers in each node, the central flower is inserted above the two lateral ones.

Salicornia perennis **P. Mill.**
Syn. *Sarcocornia perennis* **(P. Mill.) A.J. Scott**
glasswort
Native Range: East and west coasts of North America, Asia, Europe, Africa

S. perennis (*shown in photo*) looks very similar to *S. bigelovii* except that it is a woody perennial instead of an annual herb. The three flowers at an internode are inserted on the same level. Both species are edible. The ashes are rich in potash.

Wetlands 39

AIZOACEAE

Sesuvium portulacastrum (L.) L.

verdolaga, seaside purslane

Native Range: Tropical and subtropical worldwide seashores

Verdolaga is a fleshy spreading, low-growing herb. The leaves (5 cm long) are linear with smooth edges that are opposite on the stem. The stems are reddish and branched. The star-shaped pink flowers (1 cm wide) grow on a stem arising from leaf axils.

The edible verdolaga flowers year around and prevents erosion on the seashore. The common name "verdolaga" is also used for *Portulaca oleraca*, which has a yellow flower and is described in Chapter VI Roadsides and Open Areas.

40   Plants of Caye Caulker

CHENOPODIACEAE
Suaeda linearis (Elliott) Moquin-Tandon
seepweed

Native Range: Seashores from Maine to Mexico, Central America, Caribbean

Seepweed is an erect herb (20-90 cm) that is perennial in Caye Caulker, but annual in cold regions. The stems are green to reddish, branched at the slightly woody base. The leaves are narrow, linear to (7-50 mm long by 2 mm wide) with pointed ends. The flowers are very small (1-3 mm diameter), irregular, in densely branched compound spikes.

Seepweed grows in coastal wet, salty, sunny areas.

SURIANACEAE

***Suriana maritima* L.**

bay cedar

Native Range: Seashores from Florida to Brazil

Bay cedar is an evergreen shrub (to 3 m) that is multi-stemmed and spreads in older plants. The gray-green leaves (5 cm long) are spatulate and somewhat succulent. Foliage has a mild cedar-like fragrance. The sessile leaves are crowded at the end of the branch. The yellow flowers (1 cm across) are in a few inflorescences almost hidden among the leaves. The fruit is a dry drupe, surrounded by the sepals.

Bay cedar is a very salt tolerant shrub that can grow in partial shade, but it is uncommon on Caye Caulker. The local name "bay cedar" is also used on the Belize mainland for a tree, *Guazuma ulmifolia* Lam.

TYPHACEAE

***Typha domingensis* Pers.**

cattail, southern cattail

Native Range: Tropics and subtropics America, now worldwide.

Cattails grow to three meters. The flowering shoots become two centimeters thick in the middle. The leaves emerge from a flowering shoot below or just above the water level. The male (staminate) flowers are near the apex of stems and the female (pistillate) flowers below them, separated by as much as eight centimeters of bare stem.

Cattails grow in open areas in seasonal lagoons. There are two other species of *Typha* that grow in North American, but they are very difficult to distinguish because they are quite variable and have microscopic flowers.

CHAPTER V
Littoral Forest

The littoral forest habitat is on the dry sand ridges of the Caye. The high sand ridge, up to two meters above sea level, is on the east (windward) side of the Caye. The low sand ridges, up to half a meter high, can be found between some seasonal lagoons south of the airstrip. Before development the high relatively dry, sandy ridge of Caye Caulker was covered with a littoral forest, so called because it is close to the sea and the vegetation is influenced by and adapted to occasional flooding and salt-laden wind. The trees and shrubs are salt resistant and smaller than they would be on the mainland where the wind is less severe and not salty. Many plants have been introduced into this habitat, both Belizean and foreign.

Most of the littoral forest is gone, fallen victim to the need for housing and commercial development. On the south island only pockets of the original littoral forest remain, such as, the CCBTIA Mini-reserve near the airstrip. On the north island the Caye Caulker Forest Reserve at the extreme northern end protects nearly 100 acres of littoral forest. In scattered private gardens some of the original forest remains, mostly trees valuable for fruit or shade. Because the owners of these private yards often planted small trees or cuttings acquired on the mainland, it is not always possible to determine whether a shrub or tree is part of the original forest.

Bursera simaruba (L.) **Sarg.**
gumbolimbo

BURSERACEAE

Native Range: tropical Central and South America

Gumbolimbo is a medium-sized tree with reddish peeling bark. Under optimal growing conditions the trunk can reach one meter in diameter, but on Caye Caulker the trunk rarely exceeds 30 centimeters. The leaves are pinnately compound with three to seven elliptic leaflets (7 cm long). The leaves turn brown and drop in the wet season (December-February). New leaves are formed very soon thereafter. The flowers are in long clusters at the top of the tree. The fruit (1.5 cm long) is a capsule containing one seed with a red covering.

Birds eat the fruit. The bark is an anti-dote to the poison of the chechem (*Metopium brownei*), which grows only a short distance away in the same forest.

RUBIACEAE

Chiococca alba (L.) Hitchc.

milkberry

Native Range: North, Central and South America, Mexico, and Caribbean

Milkberry is a forest shrub or climbing plant growing to two meters. The leaves are opposite, simple, and ovate with smooth edges on short petioles (10 cm long). The funnel-shaped whitish to pale yellow flowers are in short axillary racemes. The fruits (5 mm across) are round white berries with a black point at the apex.

46 Plants of Caye Caulker

CHRYSOBALANACEAE
Chrysobalanus icaco L.
cocoplum, hicaco
Native Range: Florida, Mexico, Caribbean, Central and South America, introduced Pacific and Africa

Cocoplum is a rambling shrub or small tree growing to six meters. The glossy, leathery leaves (5 cm long) are nearly round, with smooth edges and a small notch at the apex. They are attached by a short petiole in an alternate configuration. The flowers are white, arising from the base of the leaf or at the ends of branches. The round fruit (5 cm long) has white flesh outside a stony seed. The outer skin of the fruit comes in three colours: white, pink and purple. The purple variety is called "black cocoplum."

The fruit is edible and used for preserves. The Black Catbird, (*Melanoptila glabrirostris*) feeds on the fruit. The Spanish name for the island, Cayo Hicaco, means "island of the cocoplum". For early settlers the cocoplum on Caye Caulker indicated the presence of fresh ground water.

Littoral Forest   47

POLYGONACEAE

Coccoloba uvifera (L.) L.

sea grape

Native Range: Mexico and Caribbean, Central and South America, now in Florida,

Sea grape is an evergreen shrub or tree that grows to ten meters. The bark is splotchy grey. The large leaves (20 cm wide) are leathery, round with a heart-shaped base and a slightly indented tip. They turn red before dropping. The white flowers are on long spikes (20-25 cm). Female trees produce round purplish fruits with flesh outside a central stony seed.

Sea grapes grow close to the sea. The fruits are edible when ripe and also made into wine or preserves. The wood is extremely hard and used for firewood.

48 Plants of Caye Caulker

BORAGINACEAE

***Cordia sebestena* L.**
zericote, geiger tree

Native Range: Central and South America, now cultivated elsewhere in Tropics

Zericote is a small to medium evergreen tree, growing to five meters on Caye Caulker. The ovate leaves (10-20 cm long) have wavy edges and a coarse surface like sandpaper. They are alternate. Zericote trees occasionally lose most of their leaves, but new leaves emerge within a few days. The bright orange flowers bloom year round in wide inflorescences. Fruits (3-5 cm long) have white flesh outside a stony seed. Flowers are visited by hummingbirds.

Cordia dodecandra A. DC., anacahuita or yellow zericote, is a similar species with bright orange flowers and large yellow edible fruit that is eaten raw in salads or stewed in syrup. The hard wood is used to make beautiful wood carvings. The leaves can be used as sandpaper. It occurs infrequently on Caye Caulker, but more frequently on the mainland.

CELASTRACEAE

Crossopetalum rhacoma **Crantz**

wild cherry

Native Range: southern Florida, Mexico, Caribbean, Central America, Venezuela

Wild cherry is an evergreen shrub that grows to three meters or more. The leaves are lanceolate to obovate with finely toothed edges growing opposite on very short petioles (1 mm long). The small, greenish-pink flowers grow from the base of the leaf. The fruit is red with a stony seed (5 mm across).

Wild cherry grows in dry soil in open areas. It has medicinal use.

RUBIACEAE
Erithalis fruticosa L.
ink plant, blossom berry, black torch
Native Range: Florida, Mexico, Central America, Caribbean

Ink plant is a shrub growing to two meters. The stiff leaves are darker green above and lighter underneath with smooth edges. They are obovate-oblanceolate and grow opposite on short petioles (5 mm). The small white star-shaped flowers (5-10 mm across) bloom year round in lateral or terminal panicles on short stems from the base of the leaves. The round black fruit (4 mm across) has five small nuts.

Ink plant grows in dry, shady areas near the beach. The wood was used for torches by the ancient Maya because the resin is flammable.

RUBIACEAE

Ernodea littoralis Sw.

yellow jugs, golden creeper

Native Range: Central America, Caribbean

Yellow jugs is an evergreen shrub (to 1 m) that forms a dense ground cover. The narrow lanceolate (5 cm long by 1 cm wide) leaves grow on short petioles (1-2 mm) with pointed tips. When grown in sunlight the leaves turn yellow, giving the plant a golden shine. The small white to pink tubular flowers bloom on short stems in leaf axils. The fruit (7 mm across) is a fleshy orange berry.

The flowers attract hummingbirds and bananaquits. It grows in full sun, tolerates drought, and can be grown in gardens if kept trimmed.

52 Plants of Caye Caulker

MORACEAE

Ficus crassinervia **Desf. ex Willd.**

fig, bearded fig, short-leafed fig, jaguey, alamo

Native Range: Tropical and sub-tropical America

The fig is a medium-sized tree (5-6 m) with aerial roots. The leaves (4-13 cm long) are elliptic to ovate, glossy, somewhat leathery, and have round tips and pointed bases. The very small flowers are hidden because they are on the inside surface of a hollow inflorescence that becomes the fig. The edible figs (1 cm diameter) are reddish when ripe. Latex is produced in all parts of the tree.

Figs are important food for fruit-eating birds, especially during migration.

VERBENACEAE

Lantana involucrata L.

sweet oregano, wild sage, cimaron

Native Range: from Florida to South America, now ornamental elsewhere

Sweet oregano is a bushy shrub (to 1.5 m). The pale green leaves (4 cm long by 2.5 cm wide) are oval–elliptic–oblong and grow opposite. The leaf edges are serrate and the surface minutely hairy. The small tubular flowers (4 mm wide) are lavender to bluish with yellow centers blooming year round in terminal dense clusters. The fruit (3-4 mm across) is a blue or purple berry.

It grows in sunny areas close to the sea. The leaves are aromatic when crushed and used as a spice or for medicinal purposes. Branches have been used as brooms to sweep the yard. In spite of the two common names, this plant is neither the commonly used oregano nor sage. Sweet oregano is important for butterflies.

ANACARDIACEAE
Metopium brownei (Jacq.) Urb.
chechem, black poisonwood
Native Range: Central America, Caribbean

Chechem is an irregularly branched tree with rough grey bark and black spots oozing resin. The glossy leaves (5-10 cm long) are generally elliptic in shape, but might range from ovate to obovate with pointed tips and arranged opposite. The leaf edges are smooth, but wavy. The minute white flowers (4 mm) are aggregated on branched stems in the tops of trees. The oblong dark red berry (5 mm across) ripens in September. The fruit is a major food source for the White-crowned Pigeon (*Columba leucocephala*).

The chechem resin causes severe skin rashes in people who are sensitive to it. It is closely related to poison oak (*Toxicodendron* sp.). The bark and leaves of the gumbolimbo (*Bursera simarruba*) are used to counter the effect of the poisonwood.

Littoral Forest   55

FABACEAE: Mimosoideae
Pithecellobium keyense **Britton**
xo-coi, buu'l che'
Native Range: Florida, Mexico, Central America, Caribbean

Xo-coi is a dense shrub growing to three meters. The compound, alternate leaves are divided into four leathery leaflets (3 cm) in two pairs, one pair on each side of the midrib with the compound leaves alternate on the stem. In each pair the leaflets are of unequal size. The pinkish flowers (2 cm across) bloom in round puffy clusters from November through February. The fruits are spirally curved pods (10 cm long) that hang in clusters. The ripe brown pods burst open, showing black beans with a fleshy red coating.

Xo-coi grows in thickets along the seashore. The red coating on seeds is edible and has medicinal use.

SAPOTACEAE

Pouteria campechiana (H.B.K.) Baehni

caramelo, canistel, egg fruit

Native Range: Mexico, Central America, now cultivated in East Asia

Caramelo is an evergreen tree that generally grows to eight meters, but sometimes to 30 meters. The glossy alternate leaves (25 cm long by 7 cm wide) are oblanceolate or lanceolate-oblong with pointed tip and tapered base that grow mostly at the ends of branches. The inflorescence is in small clusters in the leaf axil. The flowers are whitish, about one centimeter long. The edible fruit is a large smooth glossy round berry (7-12 cm long by 7 cm wide), which is yellow when ripe. The yellow flesh is soft and pasty with one to four seeds.

Littoral Forest 57

SAPOTACEAE

Sideroxylon americanum (Mill.) T.D. Penn

mul ché, mu'ul che'

Native Range: Mexico, Central America, Caribbean

Mul ché is a heavily branched evergreen shrub (3 m). The leathery leaves (3-4 cm long) are elliptic or obovate with rounded or slightly notched tips and grow in alternate configuration. The small white flowers bloom in cymose axillary clusters. The fruit is an edible round black fleshy berry (7 mm across).

Mul ché grows in sandy soil.

58 Plants of Caye Caulker

ARECACEAE
Thrinax radiata Lodd. ex Schult & Schult. f.
chit, saltwater palmetto, xa a'n che'
Native Range: Florida, Mexico, Central America, Caribbean

Chit is the only native palm on Caye Caulker. It has a slender trunk that can grow to ten meters with a diameter of 15 centimeters. The compound leaves are palmate, divided into as many as 50 leaflets. The inflorescence appears between the leaves, as much as one meter long, with small white flowers in great abundance. The fruit is a small white berry (10 mm across). The berries are an important food source for the Black Catbird (*Melanoptila glabrirostris*) and other animals.

This slow-growing palm is cut down by local people, so there are few mature trees. The straight stems are used to mark the location of lobster traps on the sea bed and for various construction purposes.

CHAPTER VI
Roadsides and Open Areas

Most of the sides of the sand roads on Caye Caulker are bordered by fences of private properties usually with houses or other buildings. A few open vacant lots remain still covered by grass and broadleaf plants. The construction and repair of the airstrip has created much disturbed land along both sides and at the east end. Many different herbs and even shrubs are found here—plants that will grow on poor soil where they are exposed to the elements.

Most of the plants we find along roadsides or elsewhere we would call weeds and not worth scrutiny by the casual observer. But there are plants with medicinal uses among these herbaceous plants. These traditions have developed over centuries of human coexistence with the plants.

Some are native to Caye Caulker; many others come from the mainland, other countries of Central America, or even from Asia. These plants grow on soil where more desirable plants do poorly and cannot compete. Some plants growing in gardens have escaped and became wild. Conversely, some wild plants have been used as ornamentals.

Weed trimmers and grass mowers put plant diversity under stress. Many properties are cleared, completely removing the vegetation cover to expose the sand. The few yards where the original vegetation has been allowed to grow were found to have a very diverse plant cover. Some large and dense stands in one year may be gone the next year with only a few remnants left. Clearing has destroyed many a stand of common plants that have since become rare.

60   Plants of Caye Caulker

ASTERACEAE
Ageratum littorale A. Gray
cloud berry, white weed
Native Range: Florida, Mexico, Central America, Caribbean

Cloud berry is a mat-forming herb (50 cm tall). The somewhat fleshy, smooth leaves (4 cm long) are deltate-ovate to oblong and opposite with toothed margins. The blue to white year-round flowers are in many flowered small clusters with no ray flowers.

It is closely related to the popular *Ageratum* bedding plant commonly found in gardens in the U.S. Cloud berry is important for butterflies.

Roadsides and Open Areas   61

ASCLEPIADACEAE
Asclepias curassavica L.
lantana, cancerillo, tropical milkweed, ba'eneno che'
Native Range: Central and South American tropics, now world wide Tropics

Lantana is an erect herb (1 m) that is often un-branched. The stem produces a milky sap. The opposite leaves (12 cm) are elliptic to lanceolate. All year round small orange flowers (5 mm across) are in umbels at the apex of the plant. The five petals are deflexed and the five filaments form a corona around the ovary. The fruit is dry (5 cm long). The seeds have long tufts of hair for wind dispersal.

Lantana is poisonous and has many medicinal uses. It is the only Belizean food plant for the Monarch butterfly (*Danaus plexippus*). The local name "lantana" is also the name of an unrelated genus with species found in Chapters V and VII.

62 Plants of Caye Caulker

ASTERACEAE

***Bidens pilosa* L.**

Shasta daisy, Spanish needles

Native Range: temperate, subtropical and tropical America, now also in Pacific and Asia

Shasta daisy is an annual or perennial branching herb (1 m). The leaves (10 cm long) are divided into three toothed lobes. The flower heads (3 cm across) are in compound cymes with several lateral branches. The disc flowers are yellow surrounded by about five irregular, broad white ray flowers.

It is a common plant with medicinal use. The common name "Spanish needles" comes from little barbs that hook onto clothes. It is important for butterflies.

NYCTAGINACEAE

Boerhavia diffusa L.

Syn. *Boerhavia coccinea* Mill.

spiderling

Native Range: worldwide tropics and sub-tropics

Spiderling is a prostrate herb with many branches spreading horizontally. The opposite leaves (2.5 cm wide) are broadly ovate to round with glandular hairs and wavy margins attached by petioles (3 cm). The year round inflorescence is in large open panicles with many branches (30 cm or more) and very small purple flowers in terminal clusters. The seeds are sticky.

Spiderling grows in dry, gravelly or sandy areas.

SCROPHULARIACEAE
Capraria biflora L.
tanchi, claudiosa, goat weed

Native Range: Subtropical and tropical America.

Tanchi is an evergreen herb or shrub (1 m tall) with erect stems, often branched, usually with leaves the entire length of the stalk. The leaves (5 cm long) are lanceolate to obovate with margins entire or serrate. The bell-shaped white flowers are five-lobed, one to three in leaf axils on short stems. The fruit is an ovoid capsule (5 mm long).

Tanchi grows in full sun on sandy soils and has medicinal use.

EUPHORBIACEAE

Chamaesyce Raf. spp.

spurge

Native Range: tropical and subtropical America, naturalized worldwide

The *Chamaesyce* genus comprises about 250 species considered by some to be part of the genus *Euphorbia*. The *Chamaesyces* look very similar with mostly technical differences. Five species have been identified on Caye Caulker. (*photos on next page*)

Three of these are closely related herbs: *C. hypericifolia, C. hyssopifolia, C. mesembrianthemifolia*. They are annual or perennial spreading and branched herbs with milk sap, usually to 25 centimeters tall. The leaves are opposite or nearly so, all lying in one flat plain. The very small flowers are unisexual and in axillary cymes in cyathia, small cups on which male flowers, each consisting of one anther, surround a single female flower of one pistil.

Chamaesyce mesembrianthemifolia **(Jacq.) Dugand** is an erect coastal plant with smooth leaf margins.

Chamaesyce hyssopifolia **(L.) Small.** and *Chamaesyce hypericifolia* **(L.) Millsp.** are easily confused as they grow side by side along roadsides. Their leaf margins are serrated or dentate. *C. hyssopifolia* has larger fruits and larger black seeds, narrower leaves, and larger stipules, while *C. hypericifolia*, has smaller fruits with brown seeds and much smaller inconspicuous red stipules.

Chamaesyce blodgettii **(Engelm ex. Hitchc.) Small.** spreads low on the ground in all directions. The young stems lack hairs.

Chamaesyce cozumelensis **Millsp.** is more erect with coarse hairs on young stems and cyathia.

66 *Plants of Caye Caulker*

Chamaesyce mesembrianthemifolia (above)
Chamaesyce blodgettii (below)

ASTERACEAE

Conyza canadensis (L.) **Cronquist**

horseweed

Native Range: North and Central America, now worldwide

Horseweed is an annual erect herb (50 cm) that is mostly branched on the upper part of the stem. The alternate leaves (10 cm long by 1.5 cm wide) are oblanceolate to linear with toothed margins. The long inflorescence is in terminal panicles. Tiny flower heads (5 mm wide) have numerous yellow disk flowers surrounded by white ray flowers that remain erect rather than spreading. The fruits are achenes with tufts of white bristles.

68   Plants of Caye Caulker

FABACEAE: Papillionoideae

Crotalaria retusa L.

rattle weed, yellow sweet pea

Native Range: Asia, Africa

Rattle weed is an erect herb (50 cm) with simple alternating leaves (4-8 cm long by 2 cm wide) that are oblanceolate to spatulate and usually retuse at tip. The yellow flowers are pea-like with red purple veins in terminal loose racemes. Thick pods (2.5-3.5 cm long) are black when ripe.

Roadsides and Open Areas 69

FABACEAE: Papillionoideae

Crotalaria verrucosa L.
beach pea, cascabel
Native Range: Asia, now pan-tropic.

Beach pea is an herb (1 m) that grows in dense stands. The alternate simple coarse leaves (5-10 cm long by 4-6 cm wide) are ovate. The flowers are mostly blue pea-like with a white area. They appear in racemes with up to ten flowers. The thick green pods (3-4 cm long) are hairy.

ASTERACEAE

Cyanthillium cinereum (L.) H. Rob.

Syn. *Vernonia cineria* (L.) Less

little ironweed

Native Range: Mexico, Central and South America, Caribbean

Little ironweed is an erect branching plant (50 cm) with simple ovate leaves (5 cm by 3 cm) along a stem with acute tips and serrate margins. The open, branched terminal inflorescences have flower heads with 20 or more purple disc flowers in bracts with short stiff hairs on short peduncles (1 cm long). There are no ray flowers.

Roadsides and Open Areas 71

FABACEAE: Papillionoideae

Desmodium incanum **DC.**

tick trefoil

Native Range: Tropical and subtropical America, now worldwide

Tick trefoil is a perennial herb with a hairy stem that is either prostrate or ascending. The alternate leaves are trifoliate with variable leaflets (4 cm long by 1.5 cm wide) are oblong or elliptic with petioles (4 cm long). Both leaf surfaces are hairy. Numerous small pink to purple flowers (5-6 mm long) are in racemose inflorescences (10 cm long) on pedicels (10 mm long). Eight-jointed pods (4 cm long) are easily split into segments with hairs with hooks that cling to clothing.

72 Plants of Caye Caulker

FABACEAE: Papillionoideae
Desmodium scorpiurus **(Sw.) Desv.**
scorpion tick trefoil
Native Range: Tropical America

Scorpion tick trefoil is a prostrate, trailing or upright, semi-woody herb with trifoliate leaves. The central leaflet (3.5 cm long by 1.5 cm wide) is longer than the two lateral ones. The leaflets are narrowly elliptic, round at both ends, and thinly hairy. The purple flowers (4 mm long) are in long racemes (10 cm). The pods (5 cm long) are constricted between joints less than two millimeters wide.

In areas with cattle it is a good forage plant for pastures.

FABACEAE: Papillionoideae
Desmodium tortuosum **(Sw.) DC.**

beggarweed, tick trefoil

Native Range: Tropical and subtropical America, now also in Australia and Pacific

Beggarweed is an erect semi-woody plant (2 m tall) with few branches. The leaves are tri-foliate, lanceolate or elliptic with the central leaflet larger than the lateral leaflets. The small purple flowers (5 mm) appear in terminal racemes on slender pedicels. The pods (3 cm long) look like a string of beads and hang downward from an upward stem.

In areas with cattle it is a good forage plant for pastures.

74   Plants of Caye Caulker

APOCYNACEAE

***Echites umbellata* Jacq.**

jasmine, rubber vine, devil's potato

Native Range: Central America

Jasmine is a trailing vine with opposite, ovate or obovate glossy leaves (5 cm long). The stems have white milksap. The white tubular flowers (4 cm across) with five sometimes slightly twisted spoon-shaped petals are in umbels with up to ten flowers. The calyx is much shorter, about one-quarter the length of the corolla tube. The fruit is a long slender pod.

Jasmine is abundant with showy flowers in the dry season. It may be poisonous. The local name "jasmine" is also used for two plants that are described in Chapter X: *Jasminum grandiflorum* and *Jasminum multiforum*.

Roadsides and Open Areas 75

GENTIANACEAE
Eustoma exaltatum (L.) Salisb.
bluebell
Native Range: North and Central America, West Indies

Bluebell is an erect, often branched herb (30 cm). The opposite leaves (2-10 cm) are elliptic to lanceolate, obtuse, both basal and along the stem. Blue-purple bell-shaped flowers appear in dry season either solitary or in small panicles with a calyx (2 cm) and an erect corolla (4 cm) that is a small tube and five lobes with dark markings in the throat. The fruit is a many-seeded capsule.

Bluebell grows on dry roadsides.

76 Plants of Caye Caulker

RUBIACEAE
Hamelia patens Jacq.
polly redhead, canan, ix ca'nan
Native Range: Caribbean, now worldwide tropics

Polly red head is an evergreen shrub (5 m). The simple opposite leaves (20 cm) are ovate, with tips acute-acuminate on petioles (3 cm). The mid-veins and edges are red. The inflorescence is in terminal cymes with tubular, red flowers (2 cm long). The fruit (1 cm across) is a berry turning from green to red to black. Flowers and fruit are both found year round.

Polly redhead is common on Caye Caulker, both wild and in gardens. It has medicinal use.

AMARYLLIDACEAE
Hymenocallis littoralis (Jacq.) Salisb.
spider lily, white lily, hook-and-eye lily, cebolla top'
Native Range: Central America

Spider lily is an herb that grows to 75 centimeters tall. The long leaves (1 m long by 7 cm wide) are lanceolate and grow from the ground up. White flowers are on one-meter-long stems in terminal umbels of 10 or more flowers. Tepals unite at the base forming a short corona (5 cm across). The long stamens protrude from the flower.

The spider lily flowers year round but each flower does not last more than two days. It is very common on Caye Caulker and important for butterflies.

CONVOLVULACEAE

Ipomoea pes-caprae (L.) R. Br.
beach morning glory
Native Range: Tropics and subtropics worldwide

Beach morning glory is a creeping vine that grows on the beach with long runners of ten meters or more. The alternate leaves (10 cm across) are orbicular to elliptic, fleshy and smooth. The bell-shaped purple flowers (5 cm across) appear in the dry season. The fruit is a round capsule (1 cm across) with a persistent calyx.

It grows on beaches where it protects against erosion.

BRASSICACEAE

Lepidium virginicum L.

pepper grass

Native Range: USA

Pepper grass (70 cm) is an erect herb that branches at the top one-third of the plant. The simple alternate basal leaves (5 cm long) have smooth surfaces, are shaped obovate to linear with serrate edges. The other leaves are pinnately lobed. The flowers are in long, terminal or lateral racemes with four white minute petals (2 mm long) and two to four stamens. The green fruit (3 mm wide) is flat with a small notch at the apex.

Lippia alba (Mill.) N.E. Br.

VERBENACEAE

oregano, sea sage, yerba che'

Native Range: Caribbean, now world wide Tropics

Sea sage is a small shrub with slender arching branches. The opposite pubescent leaves (2-4 cm long) are ovate to oblong, aromatic and slightly rough to the touch with margins serrate. One inflorescence (3 cm long) is in each leaf axil with small pink flowers.

Crushed leaves smell like *Lantana camara* and are used for tea and medicinal purposes. It is not common.

Roadsides and Open Areas 81

ONAGRACEAE

Ludwigia octovalvis (Jacq.) P.H. Raven

false primrose, willow primrose

Native Range: Americas, tropical and subtropical areas worldwide

False primrose is an herb or subshrub (2.5 m), with woody stems in older plants. The alternate leaves (17 cm long by 4 cm wide) are narrowly ovate, pointed at both ends, and sometimes pubescent on both sides. The flowers (2.5 cm across) are solitary in leaf axils with four yellow petals. The fruit is a brown cylindrical capsule (2-3 cm long) with numerous very small seeds.

82   Plants of Caye Caulker

MALVACEAE
Malvastrum corchorifolium   (Desr.) Britton ex Small
false mallow

Native Range: Florida, Mexico, Central America, Caribbean

False mallow is a perennial or annual sub-shrub (50 cm). The simple leaves are ovate with short petioles, hairy surfaces and toothed margins. The yellow flowers are axillary or in dense terminal spikes, with or without short pedicels. The staminal column is divided at the apex into short filaments. The dry fruit bursts into two halves at maturity.

ASTERACEAE

Melanthera nivea (L.) **Small**

snow square stem

Native Range: from Florida to South America, now Australia

Snow square stem is a common perennial herb (1 m). The opposite leaves are linear to ovate, often three-lobed or deltate with crenate or serrate margins on spreading stems. The round flower heads (1-2 cm across) have 30-100 white disc flowers without any ray flowers on 10-centimeter stems in axils of reduced leaves. Black anthers are prominent as black spots in each disc flower. It flowers year round.

84   Plants of Caye Caulker

Mimosa pudica L.
FABACEAE: Mimosoideae

sensitive plant, sleeper, sleepy head, durmelon, shame lady,
sleeping grass, twelve o'clock prickle, ix mu'tz

Native Range: tropical America, now worldwide.

Sensitive plant is an herb with a prickly stem that grows to one meter, but on Caye Caulker is not more than ten centimeters. The compound leaves fold when touched. The leaves are bipinnate with about 25 pairs of leaflets. The lilac flowers are in globose inflorescences (1 cm diameter).

Roadsides and Open Areas 85

Mimosa tarda **Barneby**

FABACEAE: Mimosoideae

Native Range: Central and northern South America

Like the sensitive plant, the leaves of this shrub (1 m) fold when touched. The leaves (7 cm long, 2 cm wide) are finely pinnately divided. The flowers are in pink globose inflorescences (2.5 cm). The pods are as long as the leaves or slightly longer, flat, and covered with stiff hairs along the ridges. They are purplish brown when ripe. Several pods originate from the same spot in a star-shaped configuration reflecting the shape of the inflorescence.

NYCTAGINACEAE

***Mirabilis jalapa* L.**

four o'clock, ix tze'kel ba'ak

Native Range: Mexico, introduced in Central America, South America and United States

Four o'clock is an herb with several stems that grows to one and a half meters, but is usually smaller. The opposite leaves (9 cm long) are deltate-ovate, oblong-ovate or broadly lanceolate, with pointed tips and cordate–obtuse bases on petioles (1-7 cm). The inflorescence is in clusters among small leaf-like bracts on short peduncles (5 mm). The fragrant flowers (5 cm long) are red and sometimes yellow or white. Red sepals form a tube before opening. There are no petals. Long stamens extend beyond the tube. Flowers open in the evening and wilt the next morning. The brown fruit is an egg-shaped achene enveloped by an involucrum.

Four o'clock has medicinal uses.

RUBIACEAE

Oldenlandia corymbosa L.

lang poul

Native Range: Tropical Africa, India, now pantropical

Lang poul is an annual herb (30 cm tall). The opposite leaves (35 mm long by 5 mm wide) are linear-oblong to lanceolate, acute at both ends on very short petioles. The tiny white flowers (2 mm wide) are in axillary cymes on five-millimeter pedicels. The corollas have four petals. The fruit is a capsule.

VERBENACEAE

Phyla nodiflora (L.) Greene
Syn. *Lippia nodiflora* (L.) Michx
verbena

Native Range: Tropical and subtropical America, naturalized in California and elsewhere

Verbena is a creeping herb with roots at the nodes. The leaves (4 cm long by 2 cm wide) are obovate on very short petioles (5 mm). The very small whitish-pink flowers (2 mm across) are in short compact spikes (1-2 cm long) with dark bracts.

Verbena spreads fast with runners to several meters and can form mats that help to prevent beach erosion. It is drought tolerant.

EUPHORBIACEAE

Phyllanthus amarus **Schumach.**

Phyllanthus niruri **L.**

stone breaker, chanca piedra, children weed, worm weed, leche' che'

Native Range: Neotropics, now also in India, Pacific

These two species of *Phyllanthus* are low-growing herbs (50 cm tall). The stems have many opposite, elliptic and glabrous leaves (1 cm long) on branches that are often curved. The very small flowers look like small beads. They have one male and one female flower together, without a corolla on very short pedicels (2 mm) in each leaf axil.

Both species are very similar and sometimes *P. amarus* is considered a subspecies or variety of *P. niruri*. The very small seeds of *P. amarus* (1.5 mm) have C-shaped ridges, while those of *P. niruri* have pits.

P. amarus grows on low land at elevations of 0-300 meters while *P. niruri* normally grows in the mountains at elevations of 1,000 to 2,500 meters. That *P. niruri* is found on Caye Caulker demonstrates how humans can inadvertently transport seeds from one location to another where they normally do not occur but take hold as an imported plant. Both species are widely used for a variety of diseases.

90 Plants of Caye Caulker

URTICACEAE
Pilea microphylla (L.) Liebm.
artillery plant
Native Range: Mexico to tropical South America

Artillery plant is a small annual, densely branched, prostrate herb. The opposite succulent thick leaves are oblong–obovate–suborbiculate with petioles. The leaves are in unequal pairs (2-9 mm long). The minute flowers are unisexual with no pedicel. Anthers eject pollen by force, which is why it is called "artillery plant".

Artillary plant grows on moist banks or on stone walls. It is a house plant in colder climates.

ASTERACEAE
Pluchea carolinensis (Jacq.) G. Don
Santa Maria, cough bush, cure-for-all, ix cha'al che'
Native Range: Florida to northern South America, introduced Pacific

Two related herbs are locally called "Santa Maria". *Pluchea carolinensis* is an herb or sub-shrub (2.5 m tall) with hairy stems. The leaves (20 cm long by 8 cm wide) are ovate or elliptic, rounded at the base and pointed at the tip. Margins are entire and leaves are hairy on the upper side. The pale purple flowers are in terminal or broad inflorescences with disk flowers only.

P. carolinensis is larger than the similar *P. odorata*. The leaves are slightly aromatic when crushed, while *P. odorata* is strongly aromatic. They are used for medicinal purposes and to line hen houses to control fleas.

92 Plants of Caye Caulker

ASTERACEAE

Pluchea odorata (L.) Cass.

Santa Maria, salt marsh fleabane, sweet-scent, ah cha'al che'

Native Range: Florida to northern South America, naturalized in Africa and Pacific

Two related herbs are locally called "Santa Maria". *Pluchea odorata* is an herb or sub-shrub (75 cm tall) with a somewhat hairy stem. The slightly hairy leaves (15 by 7 cm) are ovate or elliptic with serrate margins. The pink to purple flowers are in flat-topped inflorescences. The leaves are very aromatic when crushed.

P. odorata can easily be distinguished from the larger *P. carolinensis* because of its strong scent and purple flowers. *P. odorata* grows in saline soils where it is common.

P. odorata is used for medicinal purposes and to line hen houses to control fleas.

ASTERACEAE
Porophyllum punctatum (Mill.) S.F. Blake
squirrel's tail, ix tai' tai' che'
Native Range: Florida, Mexico, Central and South America

Squirrel's tail is a broad upright shrub (to 1.5 m) that is highly branched. The leaves (2-4 cm) are ovate or elliptic, with round tips, round bases, and notched edges. The configuration is variable, but the leaves are mostly opposite. Scattered tiny glands are visible on the upper side as black dots. The flower heads (1-2 cm wide) contain only pinkish-yellow disc flowers. The fruits are in a tuft of white hairs for wind dispersal.

Squirrel's tail grows in sunny areas, such as the south side of the CCBTIA Mini-Reserve.

PORTULACACEAE

***Portulaca oleracea* L.**

verdolaga, sea purslane, little hogweed

Native Range: Middle East, now worldwide.

Verdolaga is a low mat-forming herb. The fleshy alternate leaves (3 cm long) are spatulate to obovate, round at the tip. The flowers (1 cm across) are yellow. It is a variable species and on Caye Caulker a smaller, more compact form with more flowers is found together with the normal more open variety.

Verdolaga is very salt tolerant. It is edible and rich in antioxidants. It can be fed to poultry to reduce the cholesterol in eggs. The common name "verdolaga" is also used for *Sesuvium portulacastrum*, which has pink flowers and is described in Chapter IV Wetlands.

VERBENACEAE

***Priva lappulacea* (L.) Pers.**

cat's tongue, pech ma'am

Native Range: subtropical and tropical America

Cat's tongue is a hairy herb (50 cm). The opposite leaves (10 cm long) are ovate with pointed or blunt tips and toothed margins on very short petioles. The arching inflorescence (5–20 cm long) is in a terminal slender raceme. The tiny pale blue to violet flowers are tubular below petal lobes. Enclosing the fruit is the persistent inflated calyx (6 mm long by 4 mm wide) covered with hooked hairs that can cling to clothing. The fruit consists of two nutlets (3mm).

96   Plants of Caye Caulker

APOCYNACEAE
Rauvolfia tetraphylla L.
devil pepper
Native Range: tropical America

Devil pepper is a shrub (1 m) with whorls of three to five leaves, unequal in size. The leaves (10 cm long by 2 cm wide) are ovate, oblong or ovate-elliptic, broadly cuneate or rounded at the base with acute or rounded tip. Inflorescence is shorter than the leaves, lateral, unbranched with few white or yellowish flowers (2-3 mm long). The fruit is a globose drupe (1 cm across) that is red at first and then turns black.

The branches have latex. The plant is uncommon on Caye Caulker, but has medicinal use.

PHYTOLACCACEAE

Rivina humilis L.

bloodberry

Native Range: tropical and subtropical America, Africa, Pacific

Bloodberry is an herb (1.5 m). The bright green alternate leaves (10 cm long) are ovate to oblong with pointed tips and petioles. White or greenish flowers (2-3 mm) are in racemes on a ten-centimeter stem. Year round berries (3-4 mm diameter) are red with red sap.

Bloodberry is very common on Caye Caulker.

98   Plants of Caye Caulker

ACANTHACEAE
Ruellia brittoniana **Leonard.**

Syn. *Ruellia tweediana* **Griseb**

common ruellia, Mexican bluebell, Mexican petunia

Native Range: Mexico, Guatemala and Belize, now also Florida

   Common ruellia is a spreading perennial herb (1 m). The simple opposite leaves (20 cm by 1.5 cm wide) are linear-lanceolate with purple venation and serrate margins. The year-round conspicuous blue or lavender flowers are trumpet shaped in cymose panicles (3 cm across). The fruit is in a two-centimeter long capsule.

MALVACEAE

Sida acuta Burm. f. and *Sida rhombifolia* L.

chi chi be, broom weed

Native Range: pantropics

Chi chi be is a bushy herb. The leaves (3-5 cm long by 0.5 to 1.5 cm wide) are lanceolate with acute or acuminate apex, rounded base, serrate margins and petioles (5 mm long). The yellow flowers are axillary, solitary or in pairs, with pedicels (5 mm long). The flowers (1 cm across) have stamens in short tubes. The fruit is a capsule.

S. rhombifolia is very much like *S. acuta* except *S. rhombifolia* has flowers on pedicels as long as the leaves (1-3 cm). *S. acuta* spreads horizontally while *S. rhombifolia* is usually erect.

Both species have medicinal uses.

100 Plants of Caye Caulker

SOLANACEAE

Solanum americanum **Mill.**

yerba mora, nightshade, ix cha' yu'uk

Native Range; Southern United States to Paraguay, now worldwide.

Yerba mora is an herb (1 m). The alternate leaves (10 cm long by 7 cm wide) are variable with a petiole (4 cm) and toothed or wavy margins. The star-shaped flowers (1 cm wide) are white with yellow stamens. The fruit is a round black berry (5 to 10 mm wide) with many small seeds.

Although the fruit is poisonous, the leaves are used as a kitchen herb.

SOLANACEAE

Solanum donianum Walp.

potato tree

Native Range: Florida, Central America, Caribbean

The potato tree is a shrub that can grow to 2.5 meters, but on poor soil is no more than one meter tall. The alternate leaves are lanceolate with pointed tips, smooth edges and short petioles. In older plants the leaves are crowded at the top of the stem. The flowers have prominent yellow anthers with white petals. The fruit (5 mm across) is a red berry.

102 Plants of Caye Caulker

FABACEAE: Caesalpinioideae
Sophora tomentosa L.
yellow necklace, silver bush, necklace pod
Native Range: Old world tropics, now worldwide

Yellow necklace is a shrub (3 m) that is pubescent all over. The compound leaves are oddly pinnate with nine pairs of ovate to obovate leaflets and one terminal leaflet (each 2-5 cm long) covered with hairs which gives the plant a silvery appearance. The yellow flowers are in racemes (25 cm long) often standing up. The pod (15 cm long) is constricted between the seeds, causing them to look like a necklace.

The seeds are poisonous. The Yucatan Vireo (*Virio magister*) eats the budding flowers before they open, leaving only a few flowers at the apex of the raceme.

RUBIACEAE

Spermacoce verticillata **L.**

button weed, botan blanco

Native Range: tropical America

Button weed is an herb with square stems. Under favorable conditions it grows to one meter or more with woody stems, but on Caye Caulker it is usually no more than 30 centimeters tall. Opposite linear leaves (6 cm long) are in clusters at the nodes on very short pedicels. The inflorescence is in globose heads (1 cm diameter) in terminal or lateral positions with white flowers.

Button weed produces nectar for bees and has some medicinal use.

ASTERACEAE

Sphagneticola trilobata (L.) **Pruski**
Syn. *Wedelia trilobata* (L.) **Hitchc.**
rabbit's paw, u' moch atu'rich
Native Range: Tropical America and now worldwide.

Rabbit's paw is a low-growing, creeping, dense mat-forming herb that will climb on fences. The simple opposite leaves are obovate with serrate and lobed margins, usually three lobes. The year-round flower heads (2-3 cm across) have yellow ray and disc flowers.

This plant is a very common ground cover that helps to prevent beach erosion. The leaves are rich in alkaloids.

Roadsides and Open Areas 105

LOGANIACEAE

Spigelia anthelmia L.
worm bush, pinkroot
Native Range: United States to South America

Worm bush is an erect annual herb (30 cm). The simple opposite leaves are ovate with smooth margins and pointed tips. They are clustered at the tip of the stem without petioles. The flowers (1 cm across) are terminal spikes. The calyx has five teeth and the pink corolla has five lobes.

On the mainland worm bush is common, but uncommon on Caye Caulker. It is poisonous and has important medicinal properties.

106   Plants of Caye Caulker

VERBENACEAE

Stachytarpheta cayannensis **(Rich.) Vahl**

whip plant, verbena, blue vervain, ix vervaina

Native Range: tropical America, now pantropical

Whip plant is an erect perennial herb or shrub (1 m). Simple opposite leaves (12 cm long by 5 cm wide) are ovate, oblong, elliptic, or obovate with acute tip, serrate margin with or without petioles. The blue flowers are in long spikes (20 cm) without pedicels. The funnel-shaped corolla (10 mm long) has five petals, two of which are lipped, and two stamens. Only a few flowers are open at any given time. The dry fruit is enclosed by the calyx.

This species is very similar to *S. jamaicensis* (L.) Vahl. and has medicinal use.

SCROPHULARIACEAE

***Stemodia maritima* L.**

seaside twintip

Native Range: Florida, Mexico, Central America, Caribbean

Seaside twintip is an erect herb (50 cm). The opposite, simple leaves (25 mm long by 15 mm wide) are lanceolate with serrate margins. The leaves have dense sticky hairs that make them very rough to the touch. The blue or white flowers (5 mm across) are axillary without petioles with four stamens. The fruit is a capsule (3 mm long).

Seaside twintip is a source of terpenes and has medicinal use.

BORAGINACEAE

Tournefortia gnaphalodes (L.) **Roem. & Schult.**

Syn. *Heliotropium gnaphalodes* L.

sea lavender

Native Range: from Southern United States to Venezuela

Sea lavender is a grayish evergreen shrub (to 1.5 m tall by 2 m wide). The sessile leaves (4 cm long by 1 cm wide) are spathulate, alternate, and covered with dense silky hairs. New leaves are at the tip of branches and old dead leaves on lower parts of stem. The inflorescence is a scorpioid cyme with white flowers (5-6 mm wide). Fruits (1 cm across) are succulent drupes, fusing and drying to curved aggregate.

ASTERACEAE

Tridax procumbens L.

daisy, sunflower, coat buttons, tridax daisy, ba'ega che'

Native Range: tropical America, now pan-tropical

Daisy is a low creeping herb. Opposite leaves (5 cm long) are lanceolate, toothed, acute at the tips and cuneate at the bases, hairy on both sides, and on short stems. Daisy is prostrate with erect hairy flower stems (30 cm). Flower heads (3 cm wide) have five to six white ray flowers and yellow disc flowers. Flower heads often give the impression that one or more of the ray flowers are missing.

TURNERACEAE

***Turnera ulmifolia* L.**

clavo de oro, yellow alder

Native Range: Florida, tropical America introduced ornamental worldwide

Clavo de oro is a sparsely branched bush (1.5 m). The sometimes hairy, simple leaves (10 cm long) are lanceolate–oblong–ovate with pointed tips and bases, and serrate margins. Yellow flowers (3 cm across) grow from axillary buds. The fruit is an ovoid capsule.

This very common bush is used in herbal medicine and as an ornamental.

Vigna luteola (Jacq.) Benth.

FABACEAE: Papillionoideae

wild cow pea

Native Range: Africa, Asia, Philippines, naturalized Central America, Caribbean

Wild cow pea is a twining herb (50 cm tall) with trifoliate leaves. The leaflets (10 cm long by 5 cm wide) have entire margins. Yellow flowers (2 cm long) are single or in racemes with a few flowers, on peduncles (5-10 cm long) usually rising above the leaf-bearing twining stems.

Wild cow pea grows in moist, open areas in full sun.

STERCULIACEAE
Waltheria indica L.

velvet leaf, marshmallow

Native Range: Neotropics, introduced in Hawaii very early where it is considered native

Velvet leaf is an erect shrubby herb (1 m) usually with a single stem or branching at ground level. Stems and leaves are hairy, which makes them seem velvety. Alternate leaves (7 cm long) are ovate to oblong and toothed with short petioles. Small yellow flowers (5 mm) are in dense cymose clusters in leaf axils.

CHAPTER VII
Gardens

The visitor to Caye Caulker who arrives and walks or rides in a taxi on the main street will see lots of brightly coloured flowers in gardens and along fences. This is what visitors from cold climates hope to see when they venture into the tropics searching for sunshine and the blue ocean. Hibiscus, bougainvillea, oleander and ixora can be seen everywhere but these plants came from far away: hibiscus comes from Asia, bougainvillea and ixora from South America, and oleander from the Mediterranean and the Middle East.

In many gardens herbaceous ornamentals have been planted. They are not numerous but always conspicuous. All of the plants described here have been planted for their ornamental value. They do not occur in the wild except as escapees. They originate from many foreign countries and from the Belizean mainland. Their flowers and foliage make them attractive garden plants and some are used in abundance. Hibiscus and bougainvillea are so widely planted in the tropics that they are almost as synonymous with the tropics as the coconut palm is for tropical beaches.

In addition to plants used for their ornamental value and their fruit, this chapter includes kitchen herbs. The trees found in gardens are in Chapter IX. The following descriptions of herbs and shrubs of gardens are divided into sections by the colour of the flower, shown by the icon in the margin: white, yellow-orange, pink-red, blue-purple, green, and those plants for which no flowers were observed by the authors.

White

EUPHORBIACEAE

Acalypha amentacea Roxb. subsp. *wilkesiana* (**Müll. Arg.**) **Fosberg**

crinoline, rick-rack plant, Jacob's coat

Native Range: probably Pacific, now worldwide tropics

Crinoline is a densely foliated shrub (1.5 m). The green leaves with pinkish white edges (10-20 cm long) are elliptic or ovate, curled or cresty with toothed margins. The flowers are in slender whitish spikes (20 cm long and 2 cm wide).

The cultivated purple variety of this species is well known in other countries.

Gardens 115

AMARANTHACEAE  White

Alternanthera flavescens H.B.K.

joyweed

Native Range: Mexico, Caribbean, South America, introduced in Florida

Joyweed is a perennial herb with usually purple leaves and stems spreading upright or clambering. On Caye Caulker joyweed grows as a wide open shrub with long internodes and purple leaves. Opposite or alternate leaves (3 cm by 2 cm) are lanceolate or elliptic and sessile. Inflorescence (2.5 cm by 1 cm) is terminal or axillary, subglobose to cylindric. Keeled bracts are half as long as tepals. Flowers are white.

116 Plants of Caye Caulker

White

NYCTAGINACEAE
Bougainvillea x buttiana **Holttum & Standl.**
bougainvillea
Native Range: Brazil, now world wide

Bougainvillea is a shrub (5 m) with flowering vine-like, often very thorny, stems growing to ten meters long. Alternate, simple leaves (5-10 cm long) are ovate with entire margins. The small white-yellowish year-round flowers are tubular in twos or threes surrounded by three large brightly coloured bracts, varying from purple to red, yellow or white. No fruits have been observed by the authors.

Gardens 117

White

SOLANACEAE
Brugmansia suaveolens **(Humb. & Bonpl. ex Willd.) Bercht & J. Presl.**

angel's trumpet, angel's bell, tree datura, campana top'

Native Range: Tropical South America (Peru)

Angel's trumpet is a tall woody shrub (3m). Alternate simple coarse leaves (25 cm long) are ovate, smooth with entire or coarsely toothed margins. Large white flowers (20 cm long) are pendulous and trumpet-shaped.

The plant is poisonous because it contains alkaloids, especially in the seeds. It is used as a potted plant in colder climates.

118   Plants of Caye Caulker

White

EUPHORBIACEAE
Cnidoscolus chayamansa McVaugh
chaya, cha' ya
Native Range: Mexico (Yucatan Peninsula)

Chaya is an evergreen shrub or climber (3 m) with dense foliage of large leaves. The alternate leaves (10 cm or more across) are round and deeply palmate with five to seven lobes, or barely lobed with five to seven sharp points and grow on long petioles (15-30 cm). Small white flowers are in cymose flat-topped clusters on erect stems (15 cm tall) in separate male and female plants.

Chaya was a food plant for the ancient Maya. Originally it had stinging leaves, but most plants currently under cultivation only sting a little or not at all. The leaves are eaten like spinach and preferred by many because they are a source of vitamins and minerals. Raw leaves are said to be toxic. It is planted in gardens for food and grows wild as an escapee.

EUPHORBIACEAE

Codiaeum variegatum (L.) A. Juss.

garden croton

Native Range: southeast Asia, Pacific, now world wide

Garden croton is a shrub (4 m) with extremely variable foliage, from green to yellow to orange, and of many shapes and sizes. The simple alternate leaves (20 cm long and 7 cm wide) are lanceolate to ovate, glabrous and sometimes lobed. The white flowers are in small axillary racemes, but infrequently seen.

120   Plants of Caye Caulker

White

AMARYLLIDACEAE

Crinum asiaticum* L. and *Crinum sp.
milk and wine lily, red lily
Native Range: southeast Asia, introduced in other warm climates

   Milk and wine lily is a perennial herb (1 m tall). The simple leaves are in dense rosettes with many leaves at ground level or on top of a stout stem in older plants. The thick fleshy dark green leaf blades (1 m by 10 cm wide) are sessile and linear. The flowers are on top of tall thick stems (1 m) in umbels of up to 20 large flowers (15 cm wide) enclosed in two large bracts. After the flowers open, the long linear tepals curve to the outside, showing the white upper surface with purple underneath. The stamens protrude with red filaments. The plants grow large bulbs (10-20 lbs).

   Peppermint lily, another *Crinum* species with white flowers and dark pink bands on the perianth, has been seen in gardens. These plants are smaller than *C. asiaticum*. The species could not be determined. (*see photo insert*)

Gardens 121

VERBENACEAE

Lippia graveolens **H.B.K.**

oregano, Mexican oregano, ora'ego

Native Range: Central America, Texas.

Oregano is a shrub (1.5 m) with a round shape. The opposite leaves (2-4 cm long) are elliptic, ovate, or obovate with a blunt tip and round base. The leaves have hairs on both sides and dentate margins. The inflorescence is two or more flowers in each leaf axil on long stems (1 cm) with a hairy calyx (2 mm) and a slightly hairy crown (5 mm). The fruit is a small drupe.

Oregano is an important medicinal herb, as well as a culinary one. However, despite its local name, *L. graveolens* is not the commonly used oregano which is *Origanum vulgare* L.

White

MUSACEAE

Musa x paradisiaca L.
banana, plantain, platano, guinea, pra'uska'
Native Range: southeast Asia, now worldwide

The edible banana is a hybrid between two species, *Musa acuminata* Colla and *Musa balbisiana* Colla. It is a giant herb (7 m) arising from a rhizome. After flowering, the plant dies and suckers form a new "tree". The leaves (2 m long by 50 cm wide) are smooth, oblong or elliptic and grow in a spiral. The inflorescence is a terminal spike hanging down from the heart of the plant. When open the purple bract shows whorled double rows of white flowers also covered by a bract. The female flowers, 5-15 of them, occupy the lower rows. Above them are some hermaphroditic flowers and the male flowers are in the upper rows. After a few days the male and hermaphroditic flowers drop and the female flowers develop into the banana, technically a seedless berry.

Although bananas are found in Caye Caulker in gardens, they prefer humid regions with rich soil. Most plants here are the plantain or green cooking banana.

ARALIACEAE

Polyscias fruticosa **(L.) Harms**

ming tree

Native Range: probably Pacific, now cultivated in Neotropics

Ming tree is an evergreen shrub (2.5 m tall) planted for its foliage. The compound leaves have up to 15 divisions, each three to five times pinnate. Leaflets (10 cm long) are ovate to lanceolate on petioles (5 cm). The flowers are small but they appear in large, conspicuous, open umbels with brown-red branches.

Occasionally other *Polyscias* species, common on the mainland, have been seen. However, they were not in flower, which made identification impossible. Ming trees other *Polyscias* species are popular in colder climates as house plants. Ming trees are sometimes trained as bonsai.

124 Plants of Caye Caulker

White

ACANTHACEAE
Pseuderanthemum carruthersii **(Seem.) Guillaumin**
false eranthemum, ixsi'k tze'kel ba'ak

Native Range: Probably Melanesia, now cultivated worldwide

False eranthemum is a shrub (2 m) with deep purple, green or variegated, smooth leaves (5-15 cm long), which are opposite, elliptic-ovate on petioles (1-3 cm) with entire margins. The flowers (1.5 to 2 cm across) usually have four petals. The flowers are in spike-like racemes (10-15 cm). On the green and variegated forms, the flowers are white-mottled pink with a deep purple throat (*photo*). On purple forms the flowers are pink with a dark purple throat. The fruit is a capsule.

The green, variegated and purple forms are found in gardens. The flowers are quite conspicuous, especially the purple forms.

Gardens 125

CAPRIFOLIACEAE

***Sambucus mexicana* C. Presl ex DC.**

Mexican elderberry

Native Range: Southern North America, Mexico, Central America

Saúco is a shrub (4 m). The large compound opposite leaves (15 cm across) are oddly pinnate. The leaflets (5-10 cm long) are elliptic-ovate with serrate margins. The small white year-round flowers are in flat cymes (15 cm across). The fruit is a blue-black berry.

The fruit is used for jelly and wine.

White

DRACAENACEAE
Sansevieria hyacinthoides **(L.) Druce**
mother-in-law tongue, Smithsonian, snake plant, ixike'
Native Range: South Africa, introduced Mesoamerica

Mother-in-law tongue is a stemless herb. The stiff erect leaves (50 cm long) are linear-lanceolate, marbled grey or have faintly mottled bands waving across the leaves. The greenish white, tubular flowers (60 cm tall) are in clusters of three to five along stems that rise from the base in the end of rainy season. The fruits are berry-like.

This plant is closely related to and resembles *S. trifasciata*, a popular houseplant in colder regions. On Caye Caulker *S. hyacinthoides* reproduces rapidly and takes over gardens.

GOODENIACEAE

Scaevola plumieri (L.) Vahl

inkberry, fan flower

Native Range: Australia, now coastal zones of Africa and Central America.

Inkberry is an evergreen shrub (1 m high). The simple glossy, fleshy leaves (7-15 cm long) are crowded at the branch tips, elliptic, ovate–obovate with entire margin. The fan-shaped white flowers are axillary. The black fruit (1 cm across) is a drupe.

Inkberry is used as an ornamental in beach gardens because it grows in salty, sandy soil, but it does not tolerate storm surges.

128 Plants of Caye Caulker

White

Strumpfia maritima Jacq.
island romero

RUBIACEAE

Native Range: Mexico, Caribbean

Island romero is a coastal shrub with many branches with rings of stipule bases. The linear, leathery leaves (1-3 mm long) are in whorls at the end of slender branches on petioles (1 mm). Margins are entire. The inflorescence is in axillary racemes with a few white flowers (6 mm across). The corolla has five lobes. The fruit is a round drupe.

Island romero has important medicinal properties. This Caye Caulker native species is almost extinct, but is being protected in one garden.

Gardens 129

 White

COMMELINACEAE

Tradescantia spathacea Sw.

Moses-in-the-cradle, Moses-in-the-boat, spiderwort

Native Range: Mexico, Central America, Caribbean, introduced Southern United States

Moses-in-the-cradle is a perennial herb (30 cm tall). The short stem is hidden by a rosette of leaves which seem to come from the ground. The fleshy leaves (30 cm long by 8 cm wide) are green with light green stripes and purple on the underside. The inflorescence is in coiled cymes, enveloped by two bracts looking like a boat in the center of the leaf rosette. The flowers are white. It is popular in gardens and as ground cover.

130 Plants of Caye Caulker

Yellow Orange

FABACEAE: Caesalpinioideae
Caesalpinia pulcherrima **(L.) Sw.**
chac sic, pride of Barbados, ix si'nki'n
Native Range: Central America, now worldwide

Chac sic is a spreading shrub (4 m), or small tree often with thorny stems. The alternate compound leaves (40 cm long) are evenly pinnate with 8-10 pairs of ovate or elliptic leaflets (20 by 15 mm). The flowers are red with yellow margins in racemes or panicles (30 cm long). The stamens protrude from the flowers. The flowers are more regular than most flowers of this family. The fruit is a dark brown flat dry pod (15 cm long).

There is also a yellow variety (*see insert in photo*).

Yellow Orange

CANNACEAE

Canna x generalis L.H. Bail.

canna lily, platanillo

Native Range: tropical and subtropical America, now cultivated world wide

Canna lily is a large green herbaceous plant (2 m) that usually grows in clusters. The simple small banana-like leaves are oblanceolate or lanceolate with entire margins. The large year-round flowers (10 cm) vary from yellow to orange to red or red and yellow. They are asymmetric in cymes. The fruit is a capsule.

Canna lily is much hybridized and most plants are hybrids of unknown origin. Some varieties have purple leaves. The rhizome is rich in starch. Arrowroot flour is produced from a related species, *Canna edulis* Ker-Gawl.

132   Plants of Caye Caulker

Yellow Orange

HELICONIACEAE

Heliconia hirsuta L. f.

bird of paradise heliconia, u top' atzunoon

Native Range: Central America to Brazil

Bird of paradise heliconia is an herb that grows to one meter on Caye Caulker. There are three to five alternate leaves (50 cm long) on each stem. The central vein is very prominent. The flowers are on an erect stalk with six spirally arranged colourful orange bracts which enclose orange flowers. Colours may vary. The male flowers are at the apex of the spike. Flowering is in the dry season. The fruit is a drupe.

Heliconias are herbs of the humid tropical forest, spectacular in the wild and in gardens.

Gardens 133

VERBENACEAE

Lantana camara L.

wild oregano, lantana, vervain, ora'ego che'

Native Range: South and Central Americas, now world wide tropics

Wild oregano is a shrub (1 m). The simple opposite hairy leaves (10 cm long by 6 cm wide) are ovate on petioles (2 cm long) with toothed margins. The small flowers in umbels (5 cm across) are yellow-orange to red in same cluster. The colour changes as flower gets older. The fruit is a purple to blue-black drupe (5 mm across).

Some parts of this plant are poisonous. Stems and leaves are aromatic when crushed. There are many cultivated varieties.

Yellow
Orange

Senna alata (L.) Roxb.   *FABACEAE: Caesalpinioideae*

piss-a-bed, ix bra'ha

Native Range: South America or Mexico, now worldwide

Piss-a-bed is a shrub (2 m) that can grow to a small tree on the mainland. The large hairy leaves (60 cm) are even pinnately divided into eight to twenty pairs of leaflets. Leaflets (15 cm long) are ovate to oblong and obtuse at the tip. Conspicuous erect inflorescences are racemes (20 cm long) with yellow flowers. The black pods (15-20 cm long) are papery and winged.

Piss-a-bed has medicinal uses.

BIGNONIACEAE

Yellow Orange

Tecoma stans (L.) H.B.K.

trumpet bush, yellow elder

Native Range: Tropical America, naturalized in Pacific and Florida

Trumpet bush is a shrub (2 m). The opposite compound leaves are pinnately divided into three to seven leaflets that are ovate-lanceolate with pointed tips and serrate margins on short petioles. The yellow flowers (2–5 cm across) are trumpet shaped with narrow tubes, unequal petals, and four stamens. The flowers grow in showy clusters. The fruit is a capsule (10-30 cm long) with winged seeds.

136   Plants of Caye Caulker

Yellow Orange

APOCYNACEAE

Thevetia peruviana (Pers.) K. Schum.

false oleander, yellow oleander, willow, ankitz

Native Range: tropical America (Peru), now widely cultivated in the Neotropics

False oleander is an evergreen shrub (3 m). The leaves (10-15 cm long, 0.5-1 cm wide) are linear-lanceolate or lanceolate, smooth and on very short petioles (1-3 mm). Yellow-orange or yellow flowers are in cymose inflorescences (5-6 cm long) with pedicels (3-5 cm). Corollas are tubes with lobes (1-3 cm long). The fruit (2 cm) is soft and round with a very hard seed.

False oleander has medicinal use.

Gardens 137

Pink Red

EUPHORBIACEAE

Acalypha hispida **Burm. f.**
pussy tail, cattail, chenille plant, cina'an top'
Native Range: Malayan archipelago

    Pussy tails are small herbs on Caye Caulker. The simple alternate leaves (20 cm long by 10 cm wide) are ovate, pointed at the tip and usually toothed. The small female flowers are in long pink-red hanging spikes. Male flowers have not been observed.

    The milky sap is considered poisonous. In other locations pussy tail can grow as a shrub as high as five meters or as an herbaceous potted plant.

138   Plants of Caye Caulker

Pink Red

VERBENACEAE
Clerodendrum speciosissimum C. Morren
pagoda flower
Native Range: Polynesia, now widely cultivated

Pagoda flower is a shrub (2 m). The rough opposite simple leaves are ovate, cordate, densely hairy with entire or toothed margins. The bright red year-round flowers (2 cm long) are in cymes arranged in panicles (30 cm long). At first glance shrubs and inflorescences look like *Ixora coccinea*.

Pink
Red

MALVACEAE

Hibiscus rosa-sinensis L.

hibiscus, tulipan

Native Range: Tropical Asia, now worldwide Tropics

Hibiscus is an evergreen shrub (3 m). The simple alternate leaves (15 cm) are glossy, ovate–lanceolate with serrate margins and sometimes variegated. The flowers are red, but in many cultivars are white, yellow to pink, or two-toned. The flower size varies from the usual 10 centimeters to 20 centimeters wide in some cultivated hybrids. The stamens form a long tube protruding from the flower with numerous yellow anthers, the pollen-bearing part of the stamens. The style protrudes from the stamen tube and has five stigmas. Sometimes the flowers are double when stamens take the appearance of petals.

Hibiscus is the most popular garden plant in Caye Caulker and has some medicinal use.

140   Plants of Caye Caulker

Pink
Red

RUBIACEAE
Ixora coccinea L.

napoleona, ixora, jungle geranium

Native Range: India, now cultivated worldwide

Napoleona is a shrub (3 m). The opposite leathery leaves (10 cm) are smooth, oblong, sessile or nearly so, and whorled. The inflorescence is terminal in dense corymbs with up to 50 flowers. The showy year-round flowers are orange-red with cultivated varieties white, yellow or pink. A dwarf variety is sometimes found. The fruits are red or purple globose berries.

Pink
Red

EUPHORBIACEAE

Jatropha integerrima **Jacq.**

jatropha

Native Range: Caribbean, now in India, China, Malaysia, escaped in Florida

Jatropha is a large shrub or small tree (5 m) with a round shape. The leaves (10 cm or more long) are smooth, oblong-ovate and often irregularly lobed. In older parts of the tree the leaves may be very variable in size and shape. The year-round showy red flowers (3 cm) are unisexual in cymes with up to ten flowers.

142   Plants of Caye Caulker

Pink
Red

Kalanchoe blossfeldiana **Poelln.**
flaming katy

CRASSULACEAE

Native Range: Madagascar, now worldwide

Flaming katy is a low perennial herb (40 cm high). The simple alternate succulent leaves (10 cm) are elliptic–ovate–oblanceolate, glabrous with margins somewhat crenate and reddish. The red flowers (1 cm) have petals fused at the base with four lobes.

Flaming katy is a popular ornamental seen in potted plants in colder climates. Many cultivated varieties show other colours.

MALVACEAE

Pink
Red

Malvaviscus arboreus Cav.

sleeping hibiscus, Turk's cap

Native Range: tropical and subtropical America, now escapes in Pacific and New Zealand

Two species are locally called "sleeping hibiscus". Of the two, *Malvaviscus arboreus* is a shorter shrub (2 m tall) with hairy stems. The alternate hairy leaves (5-15 cm long) are ovate-suborbicular, cordate bases, acute–acuminate tips, sometimes shallowly three-lobed on petioles (10 cm) with margins serrate. *M. arboreus* has smaller red flowers than *M. penduliflorus*. They are axillary and solitary on pedicels (3 cm). The calyx is tubular and petals (3-5 cm long) do not open wide, staying almost closed. The fruit is fleshy and red.

The flowers attract hummingbirds and other birds eat the fruit.

144   Plants of Caye Caulker

Pink Red

MALVACEAE

Malvaviscus penduliflorus DC.

sleeping hibiscus, Turk's cap, ix ik tulipan

Native Range: Mexico, Central and northern South America, now worldwide in warm climates

   Two species are locally called "sleeping hibiscus" or "Turk's cap" *Malvaviscus penduliflorus* is the larger of the two species, a shrub (4 m). The hairy leaves (6-12 cm long) are lanceolate to ovate with pointed tips and toothed margin. *M. penduliflorus* has much larger red flowers than *M. arboreus*. They are solitary, never open, hang down on pedicels (3-7 cm long) and have petals 10 centimeters long. Stamens protrude from the crown.

   M. arboreus is more common in Caye Caulker gardens than *M. penduliflorus*.

Gardens 145

Pink
Red

APOCYNACEAE

Nerium oleander L.

oleander, narcisso, rosebay

Native Range: Mediterranean to Japan, now worldwide in tropical and subtropical climates

   Oleander is a shrub (5 m). The opposite gray-green leaves (20 cm long, 1-3 cm wide) are linear–oblong–lanceolate in whorls. The year-round flowers (3-5 cm across) are white, pink or red in terminal cymes. The corolla is deeply lobed and sometimes double.

   Oleander is a popular garden plant, but poisonous.

146   Plants of Caye Caulker

Pink
Red

EUPHORBIACEAE
Pedilanthus tithymaloides **(L.) Poit.**
yax xalac che, Japanese poinsettia, tza'k chu'
Native Range: Caribbean

Yax xalac che is a stiff herb (1 m). The simple alternate leaves are ovate-elliptic or obovate, often variegated, thickened and succulent with milky juice. The red flowers are in terminal clusters. The cyathium is surrounded by a two-lipped red bract (1 cm long) with a sac-like base.

This plant used to be planted near steps to keep spirits away.

Gardens   147

ROSACEAE
Pink
Red

Rosa sp.

rose

Native Range: unknown, now cultivated worldwide

Roses often are found in Caye Caulker gardens as shrubs (1.5 m tall). Usually they are not heavy bloomers and do not seem to flourish. Flowers are white, pink or red, no more than four centimeters across and double.

148   Plants of Caye Caulker

Pink
Red

SCROPHULARIACEAE
Russelia equisetiformis Schltdl. & Cham.
firecracker plant, coral plant, ix chik íchtup
Native Range: Mexico, now cultivataed elsewhere

Firecracker plant is a weeping rush-like shrub (2 m tall) with ridged stems. The stems are six- to twelve-angled. Most leaves are reduced to small scales, while other leaves (1 cm) are ovate-elliptic and dentate in whorls of three or more. The year-round red flowers (3 cm long) are narrow tubes in axillary cymes, slightly lipped, with two flowers in each inflorescence. The corolla of a flower is smooth inside.

A close relative is *R. sarmentosa*, which can be found occasionally on Caye Caulker, but distinguishes itself from the more common form by larger leaves, more than two flowers in bunches, and a corolla that is pubescent inside.

Hummingbirds visit the flowers.

AMARYLLIDACEAE

Zephyranthes lindleyana **Herb.**
Zephyranthes citrina **Baker.**
rain lily, cebollina, cebolla top'
Native Range: Mexico (*Z. lindleyana*), Central and South America (*Z.citrina*).

These two lilies are herbs that bloom after heavy rain. *Z. lindleyana* is pink while *Z. citrina* is yellow (*photo insert*). At first glance they look like crocus. The flowers are erect. Because the leaves look like grasses, the plants are hard to find when not in bloom. The perianth (3-5 cm long) is a funnel-shaped tube that consists of two sets of three petals that are not quite equal.

Rain lilies are tolerant to wet or dry weather conditions. Since they cover considerable area they make quite a show when in flower.

150   Plants of Caye Caulker

Blue Purple

APOCYNACEAE

Allamanda blanchetii **A. DC.**

purple allamanda, a'k top'

Native Range: Brazil, now worldwide

    Purple allamanda is a shrub that is either free-standing or partly climbing against fences. The simple opposite glossy leaves are elliptic-ovate with entire margins. The large, purple flowers (7-10 cm across) are bell shaped. The fruit is a dry pod. Purple allamanda looks like a large yellow allamanda (*Allamanda catharctica*).

    There are garden varieties of unknown origin with even larger purple or violet flowers.

Gardens 151

APOCYNACEAE

Catharanthus roseus **(L.) G. Don.**
periwinkle, bicaria
Native Range: Madagascar, now worldwide

Periwinkle is a perennial herb (50 cm). The simple glossy leaves (5-10 cm long) are oblong-lanceolate with entire margins. The year-round flowers (4 cm across with a 2 cm long tube) are white or violet. The fruit is a follicle (3-5 cm long).

Periwinkle is common in gardens on Caye Caulker. It has medicinal uses.

152 Plants of Caye Caulker

Blue Purple

ASCLEPIADACEAE
Cryptostegia grandiflora R. Br.
catalina, rubber vine

Native Range: Madagascar, now Pantropics

Catalina is a woody vine or shrub (3 m). The opposite leaves are elliptic or almost round (10 cm by 6 cm) glossy with acute tips and cuneate bases. The purple flowers grow in small cymes (5 cm long). The tube of the corolla is deeply five-lobed. Fruits have not been seen by the authors on Caye Caulker.

Catalina is a common ornamental.

VERBENACEAE

Duranta repens L.

dew drop, pigeon berry

Native Range: Central and South America, now world wide

Dew drop is an evergreen shrub (2 m tall) with long arching branches. Sometimes the branches are spiny. The opposite leaves (8 cm) are simple, serrate, elliptic–ovate. The fruits are yellow or orange fleshy drupes (1 cm) that hang in clusters.

Blue Purple

Lagerstroemia indica L.

LYTHRACEAE

crepe myrtle

Native Range: India, China, Japan, now widely cultivated

Crepe myrtle is a small tree or shrub. The simple opposite glossy leaves are elliptic, ovate or obovate, and smooth with entire margins. The inflorescence is a terminal panicle with purple, pink, or white flowers. The flowers have a crinkled look with spoon-shaped petals. The fruit is a brown capsule (1-2 cm diameter).

AMARANTHACEAE
Green

Amaranthus dubius Thell.

calaloo, wild spinach, ix calalo

Native Range: Caribbean, South America, now worldwide

Calaloo is an herb (1 m) with many branches. The alternate leaves (10 cm long) are ovate, cuneate at the base, acuminate – obtuse at the tip with margins entire. The upper surface of the leaves is smooth while the lower surface is hairy. They grow on petioles equal to or longer than the leaf blade. The green flowers are in axillary clusters in the lower part of the plant and branched spikes in the upper part. The seeds (1 mm long) are black and shiny. Calaloo produces many seeds.

According to one Caye Caulker citizen, it is a fairly recent introduction and unknown as recently as a generation ago. Now it is cultivated as a green vegetable and for seasoning.

156   Plants of Caye Caulker

Green

EUPHORBIACEAE
Breynia disticha J.R. Forst. & G. Forst.
snowflake plant
Native Range: Pacific, now cultivated worldwide

    Snowflake plant is a shrub (2.5 m) with dense foliage. The alternate leaves are mottled variegated, looking like they are covered with snowflakes. They are elliptic–ovate–obovate, on short stems with entire margins. New leaves are almost entirely white and pink. The greenish flowers are small with male flowers in terminal panicles and solitary female flowers. The flowers are difficult to find because they look like young leaves. Fruits have not been seen on Caye Caulker.

CHENOPODIACEAE

Chenopodium ambrosioides L.

epazote, apasote, Mexican weed, pazote'

Native Range: Mexico, Central America, Caribbean, South America

Epazote is an aromatic kitchen herb growing to 1 m. It is annual or perennial and hairy on the stem. The alternate leaves (10 cm) are lanceolate–elliptic, coarsely serrate, obtuse at the base and attenuate at the tip on short petioles. On the underside are yellow glands. The inflorescences are terminal, branched panicles. The very small greenish flowers grow on spikes out of large green bracts. The black seeds are less than 1 mm long and contain a toxic oil.

This is a kitchen herb and also has medical uses.

158   Plants of Caye Caulker

Green

CRASSULACEAE

Kalanchoe pinnata **(Lam.) Pers.**

Syn. ***Bryophyllum pinnatum*** **(Lam.) Kurz**

siempreviva, tree of life, life everlasting

Native Range: South America, India, Africa, now worldwide

Siempreviva is an herb (1 m) that is not usually branched. Some leaves are simple and opposite, while the middle leaves can be pinnately compound with three elliptic leaflets (20 cm long and 10 cm wide) with margins crenate, often producing bulbils, on petioles (10 cm). The flowers are in paniculate cymes (80 cm long) that hang down. The papery flower is a long tube-formed calyx (4 cm) that is greenish with reddish streaks. The reddish corolla (6 cm) protrude from the calyx.

Siempreviva is an important medicinal plant.

no flower

AGAVACEAE

Agave L. spp.
agave, century plant, ixike'
Native Range: North and South American Tropics

Before this project started, one agave plant had been observed flowering in an abandoned garden (*top photo*). This agave is a large perennial with a basal rosette of stiff leaves from the ground up. The leaves (1.5 m long by 10 cm wide) are long, narrow, and serrated, with a sharp tip. Agave plants flower only once and then die, but plants form suckers which form new plants after the old plant dies. The tubular or funnel-shaped flowers are on a spike or raceme on a long stalk with bracts, often several meters tall.

Two other agave species were found in Caye Caulker, but none has been seen flowering so the species could not be determined. (*cont'd next page*)

160 Plants of Caye Caulker

no flower

Smaller agaves are found in many gardens. One has white stripes on the edges of the leaves and the other has light yellow stripes. The latter is the most common. They have never been seen in flower, so the exact species could not be determined.

In Mexico, agave is used to make mezcal and tequila. The sharp tip of the leaves along with the midrib forms a needle and thread for sowing. Plants can form an impassible hedge.

ARACEAE

no flower

Alocasia macrorrhizos (L.) G. Don
coco, elephant's ear, taro, mekl
Native Range: southeast Asia, Pacific

Alocasia is a genus of 80–100 species and closely related to *Colocasia* with about 10 species. They differ in small details and both are usually referred to as "elephant's ear" because of the very large leaves (150 cm long). *A. macrorhiza* has numerous, very small flowers born on a stem called a "spadix," which is enveloped by a large bract (spathe). The fruit is a berry.

The similar *Colocasia esculenta* (L.) Schott has roots that provide taro, an important food in southeast Asia. Both are ornamentals in humid tropical climates and house plants in colder climates. Members of this family often contain calcium oxalate, which is poisonous.

162   Plants of Caye Caulker

no flower

ASPHODELACEAE

Aloe vera (L.) Burm. f.

aloe, sink-and-bible

Native Range: Atlantic Islands, Mediterranean, now cultivated in worldwide Tropics

Aloe is an herb (50 cm) with thick, succulent leaves in a basal rosette. The blue-green leaves (50 cm) are lanceolate with small spines on the margins and spotted when young. The pink red flowers (5 cm long) are tubular on panicles (1 m), but have not been observed by the authors.

Aloe is widely grown as a potted plant because it is an important medicinal plant.

ASTELIACEAE

Cordyline fruticosa **(L.) A. Chev.**

ti plant, palm lily, top' kimen

Native Range: east Asia, now world wide

Ti plant is a shrub (1.5 m). The long narrow leaves (50 cm long by 10 cm wide) have red bands, giving the plants a red appearance. Older plants are branched and leaves are concentrated near the top. Flowers have not been seen on Caye Caulker.

It is a popular garden plant because of its red-coloured leaves. It needs full sun for the colours to be bright. In colder climates it is a potted plant.

no flower

DRACAENACEAE
Dracaena fragrans **(L.) Ker-Gawl. cv Massangeana**
corn plant

Native Range: Madagascar, tropical Africa, now cultivated worldwide

Corn plant is a shrub or tree on tall stem (2 m). At the top of plant are large leaves (25 cm long and 7 cm wide) in a dense rosette, sometimes leaving most of the lower stem bare. The fragrant white-yellow flowers are in round clusters on long stems coming from the leaf rosette, but have not been observed on Caye Caulker by the authors.

This is the cultivated variety (cv) Massangeana, which has broad yellow bands in the center of the leaves. It is a popular houseplant in cold climates.

DRACEANACEAE

Dracaena reflexa **Lam.**

pleomele

Native Range: Madagascar, tropical Africa, now pantropical

Pleomele is a large shrub (10 m) with densely foliated branches. The alternate leaves (15 cm long by 3 cm wide) are lanceolate with entire margins. The white flowers are in clusters, but have not been observed on Caye Caulker by the authors.

Pleomele is a potted plant in colder climates.

no flower

Euphorbia tirucalli L.
naked lady, skeleton plant, pencil cactus, tzac ash

EUPHORBIACEAE

Native Range: tropical Africa, introduced in India and tropical America

Naked lady is a shrub (2 m) that is heavily branched, but looks leafless. The very small leaves (20 mm long by 4 mm wide) are alternate, deciduous, oblanceolate, and sessile. Photosynthesis is mostly accomplished by the long green stems. The flowers were not observed on any of the Caye Caulker plants. If present, they would be in yellow heads at the ends of branches. There is milky sap in the stems.

Although it looks like a cactus, naked lady is not a true cactus and has no thorns. It has medicinal uses.

CACTACEAE

Opuntia sp.

scoggineal, nopal, prickly pear

Native Range: North America to South America, naturalized in the Old World

The genus *Opuntia* has about 300 species. Scoggineal is a shrub-like cactus with stems divided into flattened obovate segments. The plants are covered with spines. On some plants they are very small, on others they are large. No flowers have been seen on Caye Caulker by the authors.

Several species grow in the Caribbean and in Belize on the mainland, but the plants on Caye Caulker could not be identified.

In Mexico a scale insect (*Dactylopius coccus*), lives on *Opuntia* species. Pre-hispanic weavers used a red dye, cochineal, that was extracted from the female insects.

168 Plants of Caye Caulker

no flower

Pandanus sp.
pandanus, screw pine

PANDANACEAE

Native Range: southeast Asia, now worldwide

Pandanus grows in dense clumps, often with prop roots. They become as much as five meters tall. The long linear leaves (1.5 m long by 10 cm wide) have spiny margins. This and the dense growth makes them impenetrable when planted as a hedge. Flowers have not been observed on Caye Caulker, so the species could not be identified.

Pandanus species have many uses and medicinal properties in their native land, but on Caye Caulker it is considered a noxious weed, although it is sometimes used as a hedge.

CUPRESSACEAE 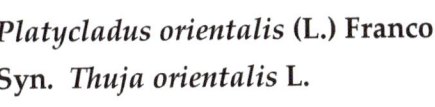 no flower

***Platycladus orientalis* (L.) Franco**
Syn. *Thuja orientalis* L.
Chinese arborvitae, biota
Native Range: China, now cultivated worldwide

Chinese arborvitae is an evergreen, bushy coniferous tree, but growing on Caye Caulker it is a shrub (2.5 m). The shape is conical when young. The leaves are very small scales on branches which are in vertical planes. The upright seed cones (2 cm long) are blue green with a waxy cover. The pollen cones are very small (3 mm long).

Although it looks like the popular thuja of American and European gardens, it differs because its seeds have no wings whereas those of thuja do have wings.

no flower

EUPHORBIACEAE
Ricinus communis L.
castor bean, higuerilla, aceite'
Native Range: Probably Africa, now worldwide

Castor bean is an erect coarse shrub (1 m). The alternate leaves (20 cm or longer) are palmately divided with five to seven lobes. The leaflets have acute tips and margins serrate. The flowers are in clusters with male and female flowers on same plant, although no flowers have been observed on Caye Caulker. The fruit is a spiny brown capsule. The seeds (1.5 cm long) are oblong and mottled with brown, black and gray. Ripe capsules burst open on the plant.

All parts of castor bean are very poisonous. It has medicinal use.

CHAPTER VIII
Trees

Plants may grow as a tree or as a shrub. We shall use the term "tree" in this book for woody plants with a clearly recognizable main trunk. A shrub would be branched low to or at the ground. Some plants that grow on Caye Caulker as shrubs can grow to big trees on the mainland.

Very big trees are rare on Caye Caulker. This is not unusual. Near the sea coast, many trees and shrubs remain small, whereas on the mainland, away from the influence of the sea, these same trees may grow to considerable height. Whether this is due to sea wind, hurricanes, poor sandy soils, salt, or a combination of them, is not always clear. Hurricanes, especially Hurricane Keith of 2000, have taken their toll. Nevertheless, there are some outstanding trees that have withstood adverse conditions and have grown to considerable size on Caye Caulker. The tallest trees are the Norfolk Island Pines (*Araucaria heterophylla*) of which there are several examples clearly visible approaching Caye Caulker from the east.

The trees described in this Chapter are neither palms, mangroves nor native trees. Instead, they are the introduced trees planted in gardens for their ornamental value or their fruit. Some are native to the mainland of Belize.

172   Plants of Caye Caulker

ANNONACEAE

Annona glabra L.

gub apple, **colcho**, pond apple

Native Range: tropical America, coastal West Africa., cultivated Asia

   The gub apple is a small tree that grows on wet soil or in swamps. The alternate, ovate to elliptic leaves (10 by 5 cm) have a round base and acute tip. The cream-coloured flowers (6 mm long) are solitary growing from leaf axils on a short thick stem. The sepals have purple spots at the base and appear in whorls of three. The flowers have many short stamens and several short pistils.

   Gub apple has two different fruits. The fruit of one pistil is round (3 cm cross) and splits open into three carpels with three or four seeds. The fruits of several pistils coalesce forming a bigger syncarp (10 cm long by 5 cm wide) with many more seeds. The pulp is pink or orange and edible. The wood of the gub apple is very light.

ARAUCARIACEAE
Araucaria heterophylla (Salisb.) Franco
Norfolk Island pine, pine
Native Range: Norfolk Island in the Southern Pacific, now worldwide

Norfolk Island pine is a tall conical tree with branches reaching up. The longest branches are at the base, tapering to the shortest at the top, giving it a triangular shape. But in 2000 Hurricane Keith stripped the branches on the west side of trees, so except for the new branches at the top, the trees have a one-sided skinny appearance.

Young leaves (1 cm long) are awl-shaped and old leaves are scale-like, overlapping and completely covering branches. Male cones (4 cm long) are in clusters and broad female cones (1-1.5 cm long) are single. Old cones can often be found on the ground.

174   Plants of Caye Caulker

MORACEAE
Artocarpus altilis (Parkinson) Fosberg
breadfruit, maxapan, masapan
Native Range: India, now worldwide Tropics

Breadfruit is a fast-growing tree that grows to 20 meters. The very large glossy evergreen leaves (75 cm by 50 cm) are deeply cut into many lobes. The foliage is very dense. Latex is abundant in all parts of the tree. The fruit (30 cm long) is cylindrical to spheroid, green and covered with four to six-sided polygons.

Some of the oldest tees in the village are breadfruit. They are planted for shade and their fruit. The whole breadfruit can be roasted, slices fried in coconut oil, or the pulp mashed and boiled as dumplings. Children used to pick the dried milky sap and chew it as chewing gum. The leaf has medicinal uses.

MALPIGHIACEAE

Byrsonima crassifolia (L.) H.B.K.

craboo, nance, acerola, chi

Native Range: tropical America

Craboo is a tree that grows to seven meters. The leaves (4-15 cm long) are opposite, clustered at end of branches, glossy on the upper side, hairy on the underside, obovate, elliptic or ovate with smooth edges. The yellow flowers are in upright terminal racemes. The popular edible fruit (1.5 cm across) is a round yellow drupe.

The tree is also planted as an ornamental.

176 Plants of Caye Caulker

Carica papaya L.

CARICACEAE

papaya, paw-paw, put'

Native Range: tropical American lowlands, now worldwide

Papaya is a small single-stem tree (5 m tall). The large toothed, many-lobed leaves (50 cm) are clustered at the top with long petioles. Male flowers are in drooping panicles at the top and female flowers are solitary in axils or below the leaves on the trunk. The fruit is a large berry, some as long as 40 centimeters with countless black seeds in a central cavity. The wood is soft with much latex.

The papaya is an important fruit with dark pink mild tasting flesh. Unripe fruit and leaves are a source of an enzyme that digests protein. It aids digestion and is used as a meat tenderizer.

FABACEAE: Caesalpinioideae
Cassia fistula L.
shower of gold, laburnum, golden shower
Native Range: India, Pakistan, now cultivated worldwide

Shower of gold is a small tree (8 m tall) with low spreading branches. The large coarse leaves (60 cm) are pinnately compound with four to eight pairs of leaflets (15 cm long). Sometimes it may temporarily lose its leaves. The yellow flowers are in long hanging racemes (40 cm long). The pistils and stamens extend from flowers curving upward. Pods (50 cm long) have not been seen on Caye Caulker.

It is an important medicinal tree as leaves, roots and seeds are used to treat diseases.

178   Plants of Caye Caulker

CASUARINACEAE
Casuarina equisetifolia L.
Australian pine, casuarina
Native Range: Australia, now worldwide

Australian pine can grow to 35 meters, but on Caye Caulker it stays much smaller because of repeated pruning and hurricanes. The leaves are minute scales on drooping branches that look like a horse's tail. The branches are in segments or internodes (9 mm long). Male spikes are four centimeters long and female cones are nine millimeters long.

They are planted frequently because they grow fast and often are pruned to keep them small or to shape them. Australian pines are an invasive species. The needles are acidic and change the pH of the soil from basic to acidic, preventing native species from growing.

Trees 179

SAPOTACEAE

Chrysophyllum cainito L.

caimito, star apple, ix cai'mita

Native Range: Central America, Caribbean, South America, cultivated Pacific

The caimito is an evergreen tree that will grow to 20 meters, but is smaller on Caye Caulker. The alternate leaves (10 cm by 5 cm) are elliptic with attenuate–obtuse tips and attenuate–cuneate bases. The upper surface is shiny green and the silky lower surface is golden to light brown. The greenish to white flowers grow in small clusters of up to 15 flowers from the leaf axils. The flowers (4 mm) have five petals and five sepals. The globose fruit (8 cm long) is purple and the cross section appears white with a purple star.

The fruit is edible.

180 Plants of Caye Caulker

RUTACEAE
Citrus aurantifolia (Christm.) Swingle
lime, li'mo'n

Native Range: tropical and subtropical Asia, now cultivated worldwide

The lime is an evergreen tree (5 m) that has thorny branches, The leaves are alternate with smooth edges. The white flowers (3 cm diameter) are strongly scented. Flowering is in late dry season. The fruit is a large green berry, globose to elongated (10 cm long).

The juice has a very sharp flavor, is rich in vitamin C, and makes delicious pies and refreshing drinks. Lime juice is also used as a cleaning agent and traditional stain-remover.

CYCADACEAE

***Cycas* sp.**

Native Range: east Asia, Australia, China, now cultivated worldwide

This tree (3 m tall) has a sturdy trunk that grows to 30 centimeters across on older plants. The smooth leaves (3 m long) are pinnately divided. The petioles (2–6 cm) have many prickles and are woolly at the base. Each leaf may have as many as 200 leaflets (30 cm long by 5 cm wide) that are linear-lanceolate to lanceolate and curved with a prominant midrib, but no lateral veins. The cylindrical male cones are in the center of the crown. The female sporophylls are loose and leafy and not arranged in a cone. The seeds (2.5 cm across) are ovoid and smooth. On Caye Caulker no plants with male cones have been observed.

Cycas is an ancient genus as shown in fossil records.

182   Plants of Caye Caulker

FABACEAE: Caesalpinioideae
***Delonix regia* (Bojer ex Hook.) Raf.**
flamboyant, mayflower, royal poinciana
Native Range: Madagascar, now cultivated worldwide

Flamboyant is a large, broad tree (10 m) with a spreading, flat crown. The leaves (60 cm long) are pinnately divided with up to 25 or more pairs of leaflets. The leaflets (5 cm) are bright green. In dry season most leaves may drop. The numerous deep orange-red flowers (5 cm wide) are in trusses. The fruits are flat pods (50 cm long by 5 cm wide).

This popular ornamental tree is magnificent in May when it flowers.

FABACEAE: *Papillionoideae*
Erythrina variegata L.
coral tree, false cassava

Native Range: tropical Asia and East Africa, now cultivated worldwide

Coral tree is a large tree characterized by large green leaves and a pale yellow midrib and veins. The branches of this tree are armed with spines. The leaves are compound with three large leaflets each over ten centimeters long. The red flowers occur in dense terminal clusters December through February. The cylindrical pods can grow to 30 centimeters.

Coral tree is fast growing and drought resistant, but uncommon on Caye Caulker. While the variety of this species that grows on Caye Caulker has yellow veins in the leaves, there are other varieties with green veins in the leaves.

184 Plants of Caye Caulker

MYRTACEAE

Eugenia uniflora L.

Surinam cherry

Native Range: South America, introduced in the Caribbean, Central America, Asia, Africa

Surinam cherry is a small evergreen tree or shrub. The opposite, simple leaves are ovate to lanceolate with pointed tips and smooth edges on very short petioles. The fragrant white flowers (15 mm across) are solitary or in small clusters of two to three flowers in leaf axils with many stamens. The edible fleshy fruit is a slightly depressed globose red berry (4 cm wide) with eight ribs.

Trees 185

MORACEAE

Ficus elastica **Roxb. ex Hornem**

rubber tree, oole' che'

Native Range: India, Indonesia, now cultivated worldwide in the Tropics

Rubber trees grow with dense foliage to five meters on the Caye. The large shiny leathery leaves (30 cm) are oblong and alternate. The flowers are inconspicuous and were not seen on Caye Caulker by the authors.

It is a popular houseplant in colder regions.

***Ficus* sp.**

MORACEAE

fig

Native Range: southeast Asia, Australia, now cultivated worldwide

These ornamental fig trees grow to three meters tall. The smooth glossy leaves (6 cm long) are alternate, oblong, lanceolate, or elliptic with tips acuminate or cuspidate and bases almost rounded. Margins are entire or nearly so. Flowers have not been seen on Caye Caulker by the authors. It may be the weeping fig, *Ficus benjamiana.*

Figs are planted in gardens because of their attractive dense foliage and are often trimmed as hedges.

FABACEAE: Papillionoideae
Gliricidia sepium (Jacq.) **Kunth ex Walp.**
madre de cacao
Native Range: Central and northern South America

Madre de cacao is a small tree (to 10 m) with alternate, pinnately compound leaves (15-30 cm long) with five to fifteen ovate leaflets. The whitish to pink flowers (about 2 cm long) are in short racemes (to 10 cm) on older branches, usually after the leaves have fallen. The flowers have a yellow blotch on the upper petal. The pods are as long as 15 centimeters.

This tree likes a humid climate with much rain. It is fast growing and will grow on sandy or rocky soils. It is used as a shade tree for cacao, which accounts for its name. It is also used for living fences, which is why it is found on the Caye. Olive-throated parakeets (*Aratinga nana*) and Bananaquits (*Coereba flaveola*) visit the flowers for nectar.

ANACARDIACEAE
Mangifera indica L.
mango
Native Range: southern Asia, now cultivated worldwide

Mango is an evergreen tree that grows to 20 meters under favorable conditions. The alternate leaves (10 to 30 cm long) are in rosettes at the ends of branches. Numerous small flowers, most of them male, grow in the dry season on profuse branched clusters (6-30 cm long). They are white to pale yellow. The fruit is a large drupe, varying in size from 6 to 25 centimeters and in weight from a few ounces to four or more pounds. The fruit colour may vary from bright yellow to green to orange-red. The seed is a flat pit in the center of the fruit.

Only a few trees have been planted in gardens. The fruit appears on the fruit stands in April. There is a story of how the mango got its name. A man had been walking a long way and was hungry. He ate one fruit from a tree and it satisfied his hunger so he could continue walking. He said that it was just enough for a "man to go".

SAPOTACEAE

Manilkara zapota (L.) P. Royen

chicle, sapodilla

Native Range: Neotropics, now cultivated in California, east Asia

Chicle is a slow-growing evergreen tree that can grow to 15 meters, but less on Caye Caulker. The wide glossy leaves (8-15 cm long, 3-4 cm wide) are spirally clustered at the tips of twigs. The small white flowers are one-centimeter wide. The edible fruit is a round brown berry (5 cm diameter). The fruit is clearly visible as it stays on the tree for 10 months.

The branches and bark contain latex that is used as chicle for chewing gum. On the mainland in northern and western Belize, chicle was a major industry until a synthetic substitute was developed.

SAPINDACEAE

Melicoccus bijugatus Jacq.

kenep, mamoncillo, Spanish lime, waya

Native Range: northern South America, Central America, now all of tropical America

The kenep tree (to 25 m) has spreading branches. The deciduous, alternate leaves are compound, with four opposite elliptic leaflets (5-12 cm long by 6 cm wide) pointed at tip. The leaf stems are frequently winged. The inflorescence is in long racemes (to 10 cm long), often clustered in terminal panicles. The fragrant white flowers have four petals and eight stamens. Male and female flowers are on separate trees. The fruits are drupes that grow in clusters.

The fruit of the variety of kenep that grows on Caye Caulker has smooth orange skin, while other varieties are green. The pleasant taste of the whitish pulp makes it a popular fruit. Humming birds visit the flowers for nectar.

RUBIACEAE

Morinda citrifolia L.

noni tree

Native Range: southeast Asia, spread to Pacific and western India, cultivated in Central and South America

Noni is a small tree (6 m). The stiff opposite leaves (25 cm long) are elliptic or oblong. The small whitish flowers are in clusters. The "fruit" (4-8 cm long) is a fleshy structure under the flowers consisting of many small fruits that have grown together into one structure called a "syncarp". The syncarp is pale yellow when ripe.

The fruit has medicinal properties and is used for a variety of ailments. It is edible but tasteless.

192   Plants of Caye Caulker

EUPHORBIACEAE
Phyllanthus acidus (L.) Skeels
grossea, whitty whitty, Tahitian gooseberry
Native Range: Madagascar, cultivated or escaped in tropical America and Pacific

Grossea is a small tree that grows to eight meters. The alternate, almost opposite, leaves (2-8 cm long, 1-4 cm wide) are ovate or elliptic with bases obtuse or cordate and pointed tips on very short petioles (2 mm), which look like leaflets of a large pinnate leaf. The racemose inflorescence is a long panicle (10-12 cm). The small flowers are simple, the male flowers with a red calyx and the female flowers with a yellow calyx. The perianth parts occur in multiples of four. The orange-yellow fruit (1 cm across) is a round fleshy berry attached to the branch.

Berries are edible, cooked or used for preserves, which is unusual for this family of plants. They are a source of vitamin C.

APOCYNACEAE

Plumeria obtusa L.

Plumeria rubra L.

frangipani

Native Range: Central and South America, now worldwide

Frangipani is a small tree (to 5 m) that is coarsely branched with latex in the branches. The long shiny, bright green leaves (20 cm) are alternate, obovate to oblong-obovate on petioles (2-4 cm). The tips of leaves are obtuse. The mildly fragrant flowers (to 7 cm across) are white with a yellow heart. They grow in terminal clusters with up to ten flowers, usually when the plant has lost leaves at the end of the rainy season.

P. rubra is another frangipani species with red or pink flowers. It has matte leaves with pointed tips. It is rare on Caye Caulker.

194 Plants of Caye Caulker

BOMBACACEAE
Pseudobombax ellipticum (H.B.K.) Dugand
Pseudobombax ellipticoideum A. Robyns
amapola, shaving brush tree
Native Range: Mexico, Central America, Caribbean

Amapola is the local name for two species that are irregularly branched trees or shrubs that can reach ten meters. They have palmately divided leaves (30 cm) with three–five leaflets. Most of the leaves are lost in the dry season. These two species are very similar but differ in the size of the flowers. Those of *P. ellipticum* are 10–18 cm long and those of *P. ellipticideum* are 7.5–10 cm long. The flowers are pink or white. In March before the leaves reappear, the flowers bloom with the appearance of a shaving brush because of the numerous protruding stamens and the curved back petals.

Amapola grows on poor soils and needs full sun. It has some medicinal value.

MYRTACEAE

Psidium guajava L.

guava, guayaba, p'ta

Native Range: tropical America, now cultivated worldwide

Guava is a small evergreen tree or shrub with scale bark. The simple opposite leaves (15 cm long by 5 cm wide) are ovate-elliptic or oblong-elliptic, pubescent below and on short petioles. The margins are entire. The white fragrant flowers (1.5 cm across) are single or a few in leaf axils with many stamens. The fruit is an oval or pear-shaped berry (10 cm long) which is green and turns yellow when mature.

The yellow flesh is rich in vitamin C.

STRELITZIACEAE
Ravenala madagascariensis Sonn.
traveler's palm, wayfarer's palm
Native Range: Madagascar, now grown worldwide

Traveler's palm is not related to the banana or palms, but to the bird of paradise. The leaves (3 m long and 50 cm wide) are on a petiole as long as the blade. Leaves are two-ranked, which gives the fan-like appearance. The inflorescences are enveloped by a huge (30 cm or more) bract with numerous white bisexual flowers. Flowering is year round but not seen very often because it occurs high up, originating between the leaves. The fruit is a capsule (8 cm). The seed is covered with a blue aril.

The leaf sheathes contain rain water which is useful in times of emergency, hence the name.

ARALIACEAE

Schefflera sp.

schefflera, umbrella tree

Native Range: Indonesia, Australia, New Guinea, now cultivated worldwide

Schefflera is a genus of small trees that grow on Caye Caulker to about five meters with straight stems and few branches. The large leaves (50 cm or more) are palmately compound, with 8 to 15 leaflets, in a whorl at the end of long petioles. The dark green leaflets (10-20 cm long) are elliptic-ovate. The small red flowers are in small heads densely arranged on long stems. Several of these stems radiate from a central axis among the foliage. The fruit is a dark drupe.

Schefflera flowers at the end of the rainy season, but is not common on Caye Caulker. The best known species of *Schefflera* is *S. actinophylla* (Endsl.) Harms (Syn. *Brassaia actinophylla* Endl.), a houseplant growing in pots in colder climates.

198   Plants of Caye Caulker

FABACEAE: Papillionoideae
Sesbania grandiflora (L.) **Pers.**
flamingo beak tree, parrot beak tree
Native Range: Malaysia, naturalized Mexico to South America

Flamingo beak is a small fast-growing, short-lived tree (to 5 m) that is sparsely branched. The leaves (30 cm) are pinnately compound with 20–40 pairs of leaflets. The large whitish or pink flowers (6-10 cm) are axillary racemes. The pods are thin and 20 centimeters long. The leaves drop in a drought. Flowers appear all year.

Flamingo beak tree has recently been imported to Caye Caulker, probably from Mexico. It can become invasive.

BIGNONIACEAE

Spathodea campanulata **P. Beauv.**

African tulip tree

Native Range: Africa, now worldwide naturalized and cultivated

The African tulip is a large tree (25 m) with odd pinnately compound opposite leaves (50 cm long) with 11–15 elliptic or oblong leaflets. The bright red flowers are on long racemes (10 cm). Each flower opens like crinkled bells, with yellow throats, frilly yellow edges and four stamens in the center. The fruits (25 cm long) are curved brown pods, with numerous paper thin seeds.

200   Plants of Caye Caulker

ANACARDIACEAE
Spondias dulcis Sol. ex Parkinson
Syn. *Spondias cytheria* Sonn.
golden plum, ambarella, Otaheite apple, golden apple, hog plum, a'bl po'k
Native Range: Polynesia, now worldwide tropics.

    Golden plum is an upright tree (12 m) with a straight trunk and smooth grey bark. The compound large leaves (60 cm) are odd pinnately divided with up to 25 leaflets. The glossy leaflets (10 cm) are lanceolate, elliptic or obovate and slightly toothed toward the apex. The leaves drop when cool weather starts. Inflorescences are large terminal panicles with white flowers. The fruit is an oval drupe (5-10 cm long) with a rough skin that hangs down from long stalks in bunches of up to 10 drupes. The pit is fibrous with tough spines into the yellow flesh, making eating or slicing difficult.

ANACARDIACEAE

Spondias purpurea L.

May plum, red mombin, jocote

Native Range: Central and South America, now worldwide in the Tropics

The May plum is a small, spreading tree or large shrub (15 m). The compound leaves (25 cm) are oddly pinnate with up to 19 almost sessile leaflets. The leaflets (4 cm) are obovate to elliptic and have a finely toothed apex. The inflorescence is a small hairy panicle with a short stem, growing in the axil of woody branches. The reddish flowers appear before the leaves. The fruit is in small clusters of two or three on very short stems and is often knobby at the apex. The edible fruit is a yellow to red or purple, ovoid or oblong drupe (5 cm) with a glossy skin. The seed is large and tough in texture. The edible flesh is yellow. The flesh is aromatic and acidic.

This species is very variable and occurs in a number of varieties. The tree is widely cultivated for its fruit.

202 Plants of Caye Caulker

MYRTACEAE

Syzygium cumini (L.) Skeels

blackberry, jambolan, java plum

Native range: Originally from China and East Indies, now in world wide tropics.

Blackberry is a small evergreen tree (6 to 20 m tall). The opposite, leathery, glossy leaves (10 cm long by 6 cm wide) are elliptic with cuneate bases and cuspidate tips on petioles (2 cm). The leaf surfaces are covered with small glands. Inflorescences are on axillary branches in cymes. The bisexual flowers (1 cm) are white to pinkish with four petals that are grown together into a cap. The calyx is very small and cuplike. Stamens are numerous. The fruit (2 cm) is an oval black berry with usually one seed.

The fruit is edible and used for making wine and jelly.

BIGNONIACEAE

Tabebuia rosea **(Berthol.) DC.**

mayflower tree, pink poui, flor de mayo

Native Range: Neotropics

Mayflower trees (up to 7 m) have palmate compound leaves (to 30 cm long) with five leaflets. The central leaflet is the longest and the two lowest leaflets are the smallest. In drought the tree may lose its leaves. The pink flowers appear in early dry season, in trusses with ten or more flowers (3 cm wide), each with a yellow heart. In some years flowers are profuse, in others they are few. The fruit is a capsule (25 cm long). Seeds have wings.

In the flowering season the mayflower trees in Belize City are spectacular.

FABACEAE: Caesalpinioideae

Tamarindus indica L.

tamarind

Native Range: east Africa, now worldwide Tropics

Tamarind is an evergreen tree (30 m). Young branches are pubescent. The leaves (16 cm long) are evenly pinnate with 10–40 leaflets (3 cm long by 1 cm wide) that are oblong, rounded at the apex, rounded and asymmetric at the base. The flowers are in racemes (20 cm long) with a pubescent stem. The flowers have red sepals that are yellow inside and petals (13 mm long) that are yellow with red veins. The thick pods (14 cm long, usually less, 3 cm across) are curved and covered with brown scuff and contain one to ten brown seeds.

The edible pods are used for culinary and medicinal purposes.

COMBRETACEAE

Terminalia catappa L.

hamans, Indian almond tree

Native Range: India, Australia, now worldwide Tropics

Hamans is a medium to large evergreen tree that branches horizontally, giving it a layered look and making it easily pruned to an umbrella shape.

The large leathery glossy dark green leaves (40 cm long) are obovate and become reddish before dropping. The very small greenish or white flowers (2 mm wide) are in spikes (20 cm). The fleshy fruit is a oval nut (5 cm) with a seed.

This tree is planted for shade, but also grows wild. It tolerates the wind and salt spray from the sea. Both the fleshy fruit and the seed are edible.

RUTACEAE

***Triphasia trifolia* (Burm. f.) P. Wilson**

limonaria, lime berry

Native Range: southeast Asia, Malaysia, now widely cultivated and naturalized

    Limonaria is a large shrub (to 3 m). The dark green glossy leaves are trifoliate with leaflets (4 cm long by 2 cm wide) on short petioles. There are one to three white flowers in leaf axils with three petals (1.5 cm long) and six stamens. The fruit is an edible ovoid red berry (1.5 cm long).

RHAMNACEAE

Ziziphus mauritiana **Lam.**

governor's plum, Indian plum, Indian cherry, Indian jujube

Native Range: India, now naturalized in Tropics

Governor's plum is a large shrub or a small tree. The alternate leaves (6 cm long by 4 cm wide) are ovate or orbicular with round tips on petioles (10 mm long). The leaves are glossy green on the upper side and pale yellow covered with fine hairs on underside with entire or almost entire margins. The whitish flowers (5mm) are in axillary cymes on short pedicels. The orange or brown fruit (2 cm wide) is a round edible drupe.

Governor's plum is planted in gardens for its fruit and is very common because it grows as an escapee.

208   Plants of Caye Caulker

Lionel Heredia (Chocolate) and Jacob Rietsema with the *Cycas*.

CHAPTER IX
Palms

Palms are flowering plants usually with a single trunk or stem and a crown of large divided leaves. Although about 900 palm species have been described for the American hemisphere, only six species are found on Caye Caulker and only one of these six is a native of the island. The only native palm is a small tree, the chit or saltwater palmetto (*Thrinax radiata*), which is described with the other native species in Chapter V Littoral Forest. The coconut palm (*Cocos nucifera*), which is the most conspicuous, is not a native of Caye Caulker but originates in Asia.

210 Plants of Caye Caulker

ARECACEAE

Chamaedorea seifrizii **Burret**
xate, bamboo palm, reed palm
Native Range: Mexico, Central America, now worldwide

Xate is an exception to the general description of palms having a single stem with a crown of leavess. It is a small palm with stems in clusters. The leaves are pinnately divided into segments about 20 centimeters long. The inflorescence is axillary below the crown of the leaf. The sessile flowers are unisexual, that is, male and female flowers are on different plants. The round black fruit (8 mm across) is a drupe, which is eaten by birds.

Xate normally grows in shaded, moist broadleaf forests on the mainland, but is found in gardens on Caye Caulker. Xate has great commercial value in the florist industry in developed countries. *"Xateros"* are going into protected areas on the mainland of Belize to illegally harvest xate.

ARECACEAE

Cocos nucifera L.
coconut palm, coco
Native Range: probably Malaysia, now worldwide

The coconut palm tree is so characteristic that tropical beaches without it are unthinkable. It can reach 30 meters, but the tall trees of the Jamaican Tall variety are disappearing due to the lethal yellow disease. Shorter resistant varieties are taking its place, such as Dwarf Malayan and Yellow Malayan, (*photo*).

The leaves (6 m long) are pinnately divided into many linear leaflets. The inflorescence comes out of the leaf axils, enveloped by a large bract, locally called "shooty". Female flowers are at the base, while the male flowers are at the apex. The large fruit consists of a thick fibrous layer which envelopes a hard "pericarp". The embryo inside is partly solid (coconut meat) and partly liquid (coconut water).

(*cont'd next page*)

When germinating, the rootlet breaks through one of the three dark, soft germination pores on the pericarp. Coconuts are everywhere on Caye Caulker, an integral part of life. Boys scramble up coconut trees to get the young coconuts, which are opened with a machette for everyone to drink the coconut water. Once the water is drained, the gelatinous meat is good to eat. For a boat launching, a large number of young coconuts are gathered to get enough coconut water to mix with rum for all the men providing the labour.

Coconuts are an essential part of Belizean cuisine. The meat of mature coconuts is grated, water added, and the mixture squeezed. What drips down is coconut milk, which is used in making bread, sweet pastries, rice and beans, soups, and other foods. Coconut milk is also used to make coconut oil. The water is removed by boiling and the oil remains. It takes 100 coconuts to make a gallon of oil.

Every part of the coconut is used. The leftover meat from making coconut milk is fed to pigs and chickens. The husks and shells are burned in traditional fire hearths for cooking.

ARECACEAE

Phoenix dactylifera L.

date palm

Native Range: from India to Egypt, now worldwide

The date palm is a tall palm (to 30 m). The trunk is covered with the remains of fallen leaves. Twenty or more leaves are clustered together with the central leaves ascending. The leaves (6 m long) are pinnately divided. The blue-green leaflets (palm fronds) are linear, stiff, and sharp. The male and female flowers are on different plants. Numerous small white flowers are on fleshy spikes more than one meter long that bend downward. When ripe, the fruits (date berries) are orange with a woody seed.

Date palms used to be grown on Caye Caulker for their fruit but only a few trees have survived. Date Lane, a small street not far from the Split, is a reminder of this once common palm.

214   *Plants of Caye Caulker*

ARECACEAE
Pritchardia pacifica Seem. ex H. Wendl.
thatch palm, botan
Native Range: Pacific, now worldwide

The thatch palm is a large palm (10 m tall) with fan-shaped leaves in a cluster at the top of the trunk. Young trees have very large (1 m) fan-shaped leaves on stout petioles (50 cm) on short trunks. The red to black round fruits (1 cm) grow in dense clusters between the leaves.

While the local names, "thatch palm" and "boton" are used on Caye Caulker, these names normally refer to different native palms that grow on the mainland and are used for pier posts (bots) and roofing (thatch palm).

Palms 215

ARECACEAE

Veitchia merrillii (Becc.) H.E. Moore
royal palm, Manila palm, Christmas palm
Native Range: Philippines, now worldwide

Royal palm is a small graceful palm that grows to six meters. Above the gray trunk is the green crown shaft that supports the crown of about 12 arched pinnate leaves (1.5 meters long).

At the point where the trunk ends and the crown shaft begins one-meter-long inflorescences appear with greenish-whitish flowers. When ripe, the oval fruits (3 cm long) are brilliantly red. They are drupes, that is, they contain a single stony seed.

Although *V. merrillii* is called "royal palm" locally, the much larger *Roystonea regia* is called "royal palm" elsewhere, but is not found on Caye Caulker.

216   *Plants of Caye Caulker*

Aurora Perez shows Jacob Rietsema how to reach the high governor's plums (*Ziziphus mauritiana*).

Angelica Novelo shows the purple allamanda (*Allamanda branchetii*) in her garden.

CHAPTER X
Climbing Plants

Wandering through Caye Caulker one cannot miss climbing plants, which grow on hedges and fences for support but would otherwise be a mass of twigs and branches on the ground. Under natural conditions, without man-made fences and hedges, they climb over whatever is growing near them. Some climbing plants twist their stems around branches of other plants; other climbers do so with the use of tendrils that attach themselves to stalks of their host plant. Sometimes a plant grows so fast that they simply drape themselves over the host without attaching themselves to it.

The distinction between climbing and non-climbing herbs or shrubs is artificial. Some plants will only climb against fences or other plants if available, but otherwise they grow as a shrub or an herb. If a plant is not found in this category the user should consult any of the other likely Chapters V, VI, or VII.

Several climbers are growing in the middle of the village on fences and in gardens. Most have beautiful flowers. The best known and perhaps the most popular vine is the allamanda (*Allamanda cathartica*) with large yellow trumpet-shaped flowers. Frequently allamanda is used on a trellis near an entrance to a house or garden.

APOCYNACEAE

Allamanda cathartica L.
allamanda, buttercup

Native Range: Brazil, now cultivated worldwide

Allamanda is a climbing shrub with leathery glossy, opposite leaves (6 cm long) that are elliptic to oblong with smooth edges and short petioles. The yellow flowers (7 cm across) are trumpet shaped, delicate, and fragrant. It flowers most of the year. A plant with double flowers has been seen.

Allamanda is a popular ornamental along fences and has medicinal properties.

Climbing Plants 219

POLYGONACEAE

Antigonon leptopus **Hook & Arn.**
lover's vine, Santiago, coral vine
Native Range: Mexico, now cultivated worldwide

Lover's vine is a dense, vigorous vine with tendrils. The simple leaves (10 cm long) are alternate and heart shaped (cordate), ovate or triangular. The flowers (1.5 cm wide) are bright pink in large terminal trusses. The tip of the central stem of the inflorescence ends in a tendril. This colorful plant can be very invasive, choking the plants it covers.

220   Plants of Caye Caulker

ASPARAGACEAE

Asparagus plumosus Baker

asparagus fern, sis che'

Native Range: South Africa, now cultivated in warm climates

Asparagus fern is a perennial, wiry climber with fibrous rhizomes and roots that can climb on fences or other vegetation to several meters high. The leaves are reduced to spine-like scales, less than 5 millimeters long. The branches are finely divided and arise in the axils of these scales that function as leaves. The very small white flowers are at the ends of branches on one-millimeter pedicels. The fruit is a black berry.

Asparagas fern is widely used by florists.

Climbing Plants 221

FABACEAE: Papillionoideae

***Canavalia rosea* (Sw.) DC.**

sea bean, beach bean

Native Range: pantropical

Sea bean is a coarse climbing vine that can grow on trees up to five meters or prostrate on beaches. The leaves (15 cm long) are alternate, thick, and divided into three leaflets that are elliptic to round and often retuse. Leaflets fold under the hot sun. The flowers (3 cm long) are in axillary racemes on long stems. They are pink-violet with white marks on the throat. Pods are 10-15 cm long, ridged, and woody. Young pods can be eaten.

This coarse vine can help prevent beach erosion.

222   Plants of Caye Caulker

VERBENACEAE
***Clerodendrum thomsoniae* Balf.**
glory bower, bleeding heart vine, rice and beans
Native Range: tropical West Africa, now cultivated by florists

Glory bower is a woody evergreen climbing vine with stems growing to more than three meters long. The coarse dark green leaves (10 cm long) are ovate with a rounded base, simple, opposite and have smooth edges. The inflorescences are terminal or in axillary cymes (10 cm wide). The red corolla (2 cm long) emerges from a white or purple calyx tube (1–2 cm wide). After flowering the corolla drops off and the calyx stays behind.

The purple form is the only one seen in Caye Caulker. The white form is a popular hanging pot plant in cold climates.

Climbing Plants 223

CONVOLVULACEAE

Ipomoea indica (Burm. f.) Merr.

morning glory

Native Range: South America, now pantropical

Morning glory is an annual or perennial twining herbaceous climber. The leaves (7 cm) are ovate, entire or three-lobed with pointed tips, cordate bases, and petioles (7-10 cm). The blue-purple flowers (6 cm) are solitary with pedicels (2 cm), or in cymes with peduncles (10 cm). The flowers open in the morning only. The fruit is a capsule (1 cm).

CONVOLVULACEAE

***Ipomoea triloba* L.**

little bell, potato vine

Native Range: tropical America, now worldwide

Little bell is an annual twining vine (1-3 m long) which, when growing over other plants, can overwhelm them. The axillary inflorescence has one or a few small funnel-shaped purple flowers (1.5 cm long) which close toward mid-day. The fruit is a globose capsule.

It grows on low herbaceous plants such as grasses and weeds.

OLEACEAE

Jasminum grandiflorum L.

Syn. *Jasminum officinale* L. var. *grandiflorum* (L.) Stokes

jasmine

Native Range: not known wild, probably India, now cultivated worldwide

Jasmine is a scrambling climber or shrub. The leaves (to 8 cm long) are opposite, pubescent, ovate, compound with two to four pairs of leaflets and a terminal leaflet. The terminal leaflets are pointed at both ends and the other leaves are elliptic. The flowers are in terminal or axillary cymes with three to nine flowers on peduncles (6 cm). The fragrant white flowers (4 cm across) have four to five lobes and two stamens.

The local name "jasmine" also refers to *J. multiflorum* (*next page*), as well as to *Echites umbellata*, which is in Chapter VI.

226   Plants of Caye Caulker

OLEACEAE
Jasminum multiflorum **(Burm. f.) Andrews**
jasmine, downy jasmine, star jasmine
Native Range: India, now cultivated worldwide

Downy jasmine can be a climbing shrub, or when standing alone it is a round shrub. The stem is covered with hairs. The leaves (5 cm long) are opposite, ovate, rounded at the base, pointed at the tip, and pubescent. The mildly fragrant white flowers are in small clusters (cymes). The calyx is hairy. The white tube-shaped corolla tube (1.5 cm long) is divided into seven to nine spreading oblong lobes. The soft berry-like fruit (2 cm across) is covered with long soft green spines.

The local name "jasmine" also refers to *J. grandiflorum* (*previous page*), as well as to *Echites umbellata*, which is in Chapter VI.

Climbing Plants 227

BIGNONIACEAE
Mansoa hymenaea **(DC.) A.H. Gentry**
garlic vine
Native Range: Mesoamerica, now cultivated in the Pacific and elsewhere

Garlic vine is an evergreen vine that spreads to three meters. The leaves (15 cm long) are opposite and divided into two ovate leaflets. The flowers are lavender with white throats in trusses. It flowers in February and March. The fruit is a flat pod (30 cm) that slowly ripens on the vine.

The leaves smell like garlic when crushed, so it is used in food as a substitute for garlic. It also has many medicinal uses.

228   Plants of Caye Caulker

CONVOLVULACEAE
Merremia dissecta (Jacq.) H. Hallier f.
twelve o'clock
Native Range: Caribbean, South America, naturalized worldwide

Twelve o'clock is an herbaceous twining vine. Palmately lobed leaves (10 cm) have five to nine lobes. The leaf margins are sinuate to dentate. The stem is hairy. The flowers (4 cm) are axillary and white with a purple heart. The corolla is bell shaped. The fruit (2 cm) is globose, surrounded by four calyx leaves.

CUCURBITACEAE

Momordica charantia L.

sorosi, bitter melon, bitter cucumber

Native Range: Old World tropics, now worldwide

Sorosi is a much-branched vine with many tendrils forming dense growth. The leaves (10 cm wide) are deeply cut into five lobes with pointed tips. The leaf margins are bluntly serrate. The yellow flowers (to 2 cm across) are on peduncles (5–10 cm long) with bracts in the lower half. The fruit (to 10 cm long) is ovoid, orange and warty. When ripe, it bursts open, shedding black seeds covered with a bright red arils.

Sorosi is an important medicinal herb the world over.

PASSIFLORACEAE
Passiflora suberosa L.
passionflower

Native Range: Tropical America, now also Pacific

Passionflower is a perennial vine with tendrils. The leaves (5 cm long) are mostly three-lobed with the central lobe being the largest. The solitary yellow-green flowers (2 cm) have five lanceolate sepals (10 mm). It flowers in March. The fruit (1 cm) is an egg-shaped glossy dark purple berry.

Passionflower grows on plants in open the forest or along the seashore.

VERBENACEAE

Petraea volubilis L.

queen's wreath

Native Range: tropical America, now cultivated worldwide

Queen's wreath is a woody vine with branches growing to ten meters long. The leaves (20 cm long) are smooth, elliptic and pointed at the tips and bases. The blue flowers (2 cm) are on axillary racemes (20 cm long). The flowers have five long petals united only at the base, giving them a star shape. It flowers in the dry season.

This showy plant is an infrequent ornamental.

ASTERACEAE
Pseudogynoxys chenopodioides **(H.B.K.) Cabrera**

Mexican flame vine

Native Range: Mexico, Central America

Mexican flame vine is a vigorous woody climbing vine. The leaves (12 cm long by 7 cm wide) are alternate and lanceolate–ovate with pointed tips. The leaf margins are serrate to dentate. Inflorescence is single or in a panicle. The year round flowers (5 cm) are orange. The fruit is an achene with bristles for wind dispersal.

Mexican flame vine is spectacular when in full flower, so it is used as an ornamental.

LORANTHACEAE
Struthanthus cassythoides **Millsp. ex Standl.**
mata palo, guard plant, scorn d'earth
Native Range: Mexico, Central America

Mata palo is a semi-parasitic rambling, climbing evergreen shrub with simple leaves (5 cm) that are mostly opposite, lanceolate, ovate or spathulate, and almost sessile with margin entire. The yellowish flowers are axillary in small clusters or on larger open cymose inflorescences with 16 or more flowers on opposite pedicels with one or two flowers each. The fruit is a sticky purple berry.

This plant often grows on buttonwood, which it can completely smother under rampant growth and eventually kill it. Although there was very little mata palo left after Hurricane Keith, by 2009 many buttonwood trees on the south end of the Caye have been smothered by the parasite.

234   Plants of Caye Caulker

Petrona Joseph tells Jacob Rietsema about all the plants in her garden.

Claudia Reyes telling Jacob Rietsema about her kenep (*Melicoccus bijugata*) that grew back from a root after Hurricane Keith.

CHAPTER XI
Grasses and Sedges

Sedges have edges, and rushes are round,
But grasses have nodes from their tips to the ground.

Although about 250 species of grasses and 150 species of sedges have been recorded in Belize, we have identified on Caye Caulker only 19 grasses and six sedges.

Grasses usually have leaves that are long and slender, flat or sometimes rolled. The flowering stems or culms have nodes and internodes with leaves at the nodes. The alternate leaves are in two rows (**two ranks**). The inflorescence is usually found at the apex of a stem but sometimes also in the axils of leaves as in corn (maize).

Sedges (nutgrasses) superficially look like grasses with solid triangular stems. The leaves are flat and often have very sharp edges and usually are basal. If on a stem, sedges grow in three vertical rows (**three ranks**). They almost always grow in wet places. The inflorescence is at the apex of a stem.

Grasses

Grasses are all in the family *Poaceae*, also called *Gramineae*. Grasses are usually low herbs with a round, hollow or solid stem with leaves. However, some grasses are very tall, such as, corn and sugar cane.

On the stem, also called "**culm**", leaves grow from **nodes**. The portion of the stem between nodes is the **internode**. A leaf consists of a **blade** and a lower part, the **sheath**, which envelopes the stem. On the border between the sheath and the blade is the **ligule**, a small and short, variable structure.

Grasses can grow in dense clusters, single, creeping or erect. Often they multiply by producing **stolons**, horizontal stems growing above the ground or **rhizomes**, horizontal stems growing underground, both of which send out new roots at the nodes.

The flowers grow in **inflorescences**, usually at the apex of a stem or less frequently in the axil of leaves, such as in corn. The flowers are very small and grow on short stems called "**spikelets**". A spikelet can have one or more **florets** that are surrounded by one or more very small **bracts**. When the spikelets are directly attached to the main stem, the inflorescence is called a "**spike**". When the spikelets are growing on a short stem or **pedicel**, which in turn is attached to the main stem, the inflorescence is a **raceme**. When the spikelets are growing on pedicels, which in turn are growing on longer branches, the inflorescence is a **panicle**. All of these structures are variable and important for identification.

The fruit is usually an **achene**, a seed with a fruit wall attached, or rarely a berry or a nut.

POACEAE or GRAMINEAE
Andropogon glomeratus **(Walter) Britton, Sterns & Poggenb**.
fox tail, broom grass, zu'uk

Native Range: tropical and subtropical America

Fox tail can be recognized by the tall dense inflorescence, forming big, coarse plumes that look like fox tails or brooms. It forms dense clumps, growing from underground stems (rhizomes) and becomes quite tall (2 m). The leaf blades are flat (10 mm wide).

It grows on sandy soils and is found along roadsides and vacant lots. It is very common on Caye Caulker.

POACEAE or GRAMINEAE
Cenchrus incertus **M.A. Curtis**
burr-burr, sand burr, burr grass
Native Range: tropical and subtropical America, now worldwide

Burr-burr is a common low (15 cm) annual grass with flat leaves. The ends of the flowering stems (10 cm) have spikes with many round spikelets. The spikelets have spines that attach themselves to clothes. Walking barefoot on Caye Caulker can be very painful if you step on a burr-burr.

Burr-burr grows on sandy soils in open areas.

Another *Cenchrus* species, possibly *Cenchrus echinatus* L., is less common on Caye Caulker, locally called "small burr-burr."

Grasses 239

POACEAE or *GRAMINEAE*
Cymbopogon citratus **(DC.) Stapf.**

lemon grass, sacate'

Native Range: southeast Asia, now cultivated everywhere in the Tropics

Lemon grass is one of a group of more than 50 species found in southeast Asia. It is a perennial grass about 50 centimeters tall that forms dense clumps. The flowering stems, which are seen September through November, may reach almost two meters.

Lemon grass grows in sunny, open areas and is widely used in cooking, so it is found in gardens.

240   Plants of Caye Caulker

POACEAE or *GRAMINEAE*
Cynodon dactylon (L.) Pers.
Bermuda grass

Native Range: Africa, now worldwide

Bermuda grass is a perennial, low, creeping grass with long runners rooting at the nodes, forming mats. The leaves (10 cm) are grey-green. Culms stand up to 40 centimeters tall with as many as ten erect spikes (6 cm) purple when in flower, originating at the apex.

Bermuda Grass will grow on poor soils and is drought resistant. It is very common.

Grasses 241

POACEAE or GRAMINEAE
Dactyloctenium aegyptium (L.) Willd.
crowfoot grass, Aegyptian grass
Native Range: Old World Tropics, now worldwide

Crowfoot grass is an annual, low (40 cm), creeping, mat-forming grass. The ends of the culms have six spikes radiating horizontally like a star. Each spike has two dense rows of many short spikelets hanging down like a double comb.

It is very common in open areas growing on sandy soils.

POACEAE or *GRAMINEAE*

Distichlis spicata (L.) Greene

salt grass

Native Range: North and South America, naturalized elsewhere

Salt grass is low (40 cm) with strong rhizomes that creep and form mats. The leaves (10 cm long) grow in two vertical rows (two-ranked) alternate and opposite close together on flat stems. The inflorescence with greenish-purple flowers is on spikes (8 cm long) with six-seven dense spikelets close to the stem.

Salt grass is named because it is very salt tolerant and commonly found along the water's edge.

Grasses 243

POACEAE or GRAMINEAE
Eleusine indica (L.) Gaertn.
goose grass, crabgrass
Native Range: Africa, naturalized elsewhere

Goose grass is a tall annual or perennial grass that forms clumps. The leaf blades (30 cm long by 8mm wide) are flat. At the end of the 50 cm long culms grow two to six spikes radiating from the apex of the stem, except for one which grows three centimeters below the apex. The spikelets are in two rows along one side of the stem.

It is common and grows on sandy, moist soils in open areas.

POACEAE or GRAMINEAE
Eragrostis Wolf

Eragrostis is a large genus of grasses with more than 300 species from temperate and tropical climates. The differences between species are mostly in the florets. The culms of some species grow to almost two meters while others do not exceed 20 centimeters. The inflorescences of most species are large open terminal panicles. The three species identified in Caye Caulker have the characteristic open panicles.

Eragrostis amabilis **(L.) Wight & Arn. ex Nees**
love grass

Native Range: Old World Tropics, now world wide

Eragrostis amabilis is an annual low (10 cm) grass. The inflorescence is open, diffuse, fine panicles (10 cm long) with dainty brown-yellow spikelets (1.5 mm). Love grass is drought resistant. (*top photo*)

Eragrostis elliottii **S.Watson**
blue love grass, Elliot's love grass

Native Range: tropical and subtropical Americas

Eragrostis prolifera **(SW.) Steud.**
Dominican love grass

Native Range: tropical and subtropical America, now Africa

E. elliottii and *E. prolifera* are two perennial love grasses that are much taller than *E. amabilis*. Both have stiff leaves that are rolled. *E. elliottii* grows to 80 cm. The culms (50 cm) are open and diffuse with purple spikelets (2 cm).

E. prolifera grows to 200 cm with a fringe of hairs on the branches. Inflorescence is an open panicle (17 cm long) with spikelets, each with 8-30 florets, linearly compressed to 15 mm long by 2 mm wide. (*bottom photo*)

E. prolifera grows on beaches or in brackish water, whereas *E. elliottii* grows on sandy woodlands, but on Caye Caulker they grow side by side. Dogs and cats eat this grass as a laxative.

Eragrostis amabilis (above) *Eragrostis prolifera* (below)

POACEAE or *GRAMINEAE*

***Eustachys petraea* (Sw.) Desv.**

finger grass

Native Range: tropical and subtropical Americas, introduced elsewhere

Finger grass is a tall (1 m) perennial grass with stolons, stems that grow above, but along, the ground. The culms grow horizontaly with the ends standing up (decumbent). The bluish blades are flat and wide (1 cm). The culm has four to six spikes at the apex, each with two rows of spikelets on one side of the stem.

POACEAE or GRAMINEAE
Panicum altum **Hitchc. & Chase**

pea shooter grass, panicum

Native Range: tropical and subtropical Americas

Pea shooter grass is a perennial grass with rhizomes and culms. The culms are lying down with the apex ascending to two meters from swollen nodes. The leaf blades are long (45 cm) and wide (1.5 cm). The inflorescence is an open panicle, open and ovate. The primary branch (30 cm) is upright with spreading secondary branches, each with single pinkish oval spikelets on stems.

In the past, children used the hollow stems to shoot seeds, hence the local name.

Paspalum L.
POACEAE or GRAMINEAE

Native Range: tropical and subtropical Americas

Paspalum is a group of more than 300 species mostly in warm climates. They are annual or perennial with culms reaching two meters. The species found on Caye Caulker have culms with a number of spike-like racemes arranged along main the stem, often separated two to three centimeters from each other. Three species have been identified on Caye Caulker:

Paspalum blodgettii **Chapm.**
Blodgett's crown grass

Blodgett's Crown Grass is a tall (100 cm) perennial and clump-forming grass. The base of the stem is swollen, bulb-like. The slender culms (100 cm) have no lateral branches on the stem. The leaf blades (25 cm long by 3-14 mm wide) have smooth surfaces. Three to eight racemes are spread on a central stem. Spikelets are in pairs on the spikes.

Paspalum langei **(E. Fourn.) Nash**
rusty-seed paspalum

Rusty-seed paspalum is a perennial grass without rhizomes. The leaf blades (25 cm long by 2 cm wide) are in a dense cluster near the roots. The culms (80 cm) have three or four racemes (6 cm) near the apex and several more laterally on the stem, each with four rows of spikelets closely pressed to the stem.

Grasses 249

POACEAE or *GRAMINEAE*

Paspalum virgatum L.

switchgrass, talquezal

Native Range: tropical and subtropical Americas

Switchgrass is a roobust tall (1-2 m) perennial grass with stems in dense clumps. The leaf blades are wide (2.5 cm), long and flat with small spines on the edges. The inflorescence is a nodding panicle with long racemes (15 cm), each with crowded pubescent spikelets (2.5 mm) in two rows on one side of the brown stems.

250   Plants of Caye Caulker

POACEAE or *GRAMINEAE*

Saccharum officinarum L.

sugar cane, to'

Native Range: tropical Asia, now worldwide Tropics

Wild sugar cane does not grow on Caye Caulker, so the sugar cane plants found in gardens are probably hybrids from commercial fields. The stalk (5 cm across) grows to several meters tall with long (50 cm) leaves springing along the stalks. At the apex are long (50 cm) plume-like panicles.

Sugar cane is found in gardens because the stems are chewed for their sweet taste.

POACEAE or GRAMINEAE
Spartina spartinae **(Trin.) Merr. ex Hitchc.**
cutting grass, **pincha huevo**, cordgrass, well zu'uk
Native Range: American Hemisphere

Cutting grass is a perennial grass that forms large (1 m) clumps. The long leaves are not flat like most grasses, but filiform (2–3 mm wide). The culms can reach two meters in height with numerous spikes (10–30 cm long). The spikes are pressed close to the stem.

Cutting grass is often found near water or mashes. The common name "cutting grass" is also locally used for a sedge, *Cyperus ligularis* (p. 255)

252   Plants of Caye Caulker

POACEAE
Zoysia matrella (L.) Merr. var. matrella
zoysia grass, Manila grass, Japanese carpet grass
Native Range: China, Japan, Australia, southeast Asia, cultivated worldwide warm climates.

Zoysia grass forms low dense mats and spreads by stolons. The narrow leaf-blades (8 cm long by 3 mm wide) are lanceolate. Inflorescences are on terminal spike-like racemes (4 cm long).

It hybridizes easily with other *Zoysia* species. Several varieties and cultivars are used for lawns and golf courses. This grass is salt tolerant, grows well on sandy and poor soil.

Sedges

Most sedges are grass-like herbs of worldwide distribution. Sedges, also called "nutgrasses", look like grasses but differ considerably.

Whereas a grass stem is round, a sedge stem is three-sided with the ribs that are often very sharp. Sedge stems lack nodes. Sedge leaves grow out of the stems in three vertical rows (three-ranked), in contrast to grasses that are two-ranked. While grass leaves are flat, sedge leaves are long and narrow with closed sheaths.

The brown, compact inflorescences are at the apex of leafless stems. They are subtended by several leaflike structures (bracts). The small flowers are arranged in spikelets that are also subtended by scales or bracts. The perianth is absent or reduced to bristles or scales. There are one to three stamens and one ovary with one style.

The fruit is a three-angled or lens-shaped achene that is flattened-convex or lens shaped. The seeds or nuts of sedges are edible. The sedge *Cyperus papyrus* was the source of paper in ancient times. Sedges are in the *Cyperaceae* family.

CYPERACEAE

Cladium jamaiscense **Crants**

saw grass

Native range: North, Central and South America, now worldwide.

Saw grass is a tall rhizomatous grass (1–3 m) growing in brackish or fresh water in dense stands. The stems are barely tri-angled. The leaves (1 m long) grow along the stem and are flat and about one centimeter wide with saw-toothed margins. The large inflorescences (50 cm) are either terminal or in the axils of leaves, wide open with many bracts and brown spikelets. The achenes (2 mm) are green to brown.

It is the dominant grass in the Florida Everglades.

CYPERACEAE

Cyperus ligularis L.

cutting grass, cortadera, chufa, Caribbean sedge or purple sedge

Native Range: tropical and subtropical America, Africa

Cutting grass is a clump-forming plant that grows to 125 centimeters tall. The stem is papillose or warty. The waxy, V-shaped leaves have prominent midribs, small papilla (pimples), and serrated edges. Culms have three prominent involucral bracts and three to seven, round or oblong purple spikes (10–25 mm) with many (20–80) spikelets. The achenes are round.

The common name "cutting grass" is also used locally for a grass, *Spartina spartinae* (p. 251)

256   Plants of Caye Caulker

CYPERACEAE

Cyperus polystachyos **Rottb.**

flat-spiked sedge

Native Range: Subtropical and tropical Americas, Africa, Asia

Flat-spiked sedge is a perennial clump-forming sedge with short rhizomess and short (50 cm tall) above-ground stems. Leaves have a prominent rib in the center. At the ends of culms are several ovoid, inversely conical spikes with a dense clump of spikelets. The involucral bracts are five to fifteen centimeters long. Flowers are inside half or fewer of the bracts. The achenes are flat.

Sedges 257

CYPERACEAE

Eleocharis sp.

spike grass, spike rush

Native range: worldwide in warm climates

Eleocharis is a genius of rush-like mostly aquatic plants that are very difficult to identify. They are perennial, rhizomatous plants with round stems that grow to one meter tall. The leaves are reduced to membranous sheaths at the base of the stems. The inflorescences (1 cm long by 5 mm wide) are in spikelets with acute or obtuse tips located at the apex of the stems. Flowers are brown with three yellow stamens. The achenes (2 mm long) are brown.

One species is the Chinese Water Chestnut (*Eleocharis dulcis*) which has edible tubers on the rhizomes.

Plants of Caye Caulker

CYPERACEAE

Fimbristylis cymosa R. Br.

button sedge

Native Range: North and Central America, Caribbean (Cuba)

Button sedge is a perennial. The leaves (3 mm wide) are half as long (25 cm) as the culms (50 cm). Inflorescence is in numerous small clusters of sessile spikelets (3 mm). The three bracts immediately under the inflorescence are equal to or shorter than the inflorescence itself. Button sedge has round fruits in dense clumps at the end of stems on very short branches, which are different from the flat fruits of flat sedge.

Sedges 259

CYPERACEAE
Fimbristylis spadicea (L.) Vahl
Syn. *Fimbristylis castanea* (Michaux) Vahl
chestnut sedge
Native Range: Tropical America

Chestnut sedge is a clump-forming plant 150 centimeters tall. The linear leaves have rolled edges and dark chestnut bases without hairs. The bracts are equal to or shorter than the inflorescence, which is a compound umbel. It can be recognized by chestnut brown cone-shaped spikelets at the end of branches off the main stem. Out of some of the chestnut brown bracts extend pink oblong flowers with thin white fibers.

Jacob Rietsema and Dorothy Beveridge showing the plant press in the drying cabinet at the CCBTIA Resource Center.

Celebrating the completion of the plant inventory are: (standing) Jacob Rietsema and Louis A. Aguilar; (sitting from left) Isela Marin, Dorothy Beveridge, Lydia Vega, Ernesto Marin, Jr., and Aurora Perez.

Glossary

Plant descriptions almost always contain technical terms. Their use makes it possible to avoid long wordy explanations and confusion. This requires that each term is defined and unambiguous. Below is a short explanation of the terms used. They are based on the glossary of *Hortus Third* by L.H. Bailey and E.Z. Bailey (1976), which in turn is based on *Taxonomy of Vascular Plants* by G.H.M. Lawrence (1951).

In the plant descriptions of this book the most used terms pertain to leaf shape and inflorescence, because both are obvious characteristics when an unknown plant is found and the identity sought. Leaf shapes and inflorescences are also defined by drawings in Chapter I.

A

Achene — small dry fruit with one seed and tight outer wall.
Actinomorph — radially symmetrical.
Acuminate — tapering to a sharp point with slightly concave sides.
Alternate — arranged singly and at different heights on the stem.
Androecium — the male element of a flower, one or many stamens.
Anther — the pollen bearing part of the stamen; pollen sac.
Attenuate — long, tapering to the tip or base.
Apex — the distal end of a stem or leaf, the tip.
Aril — fleshy appendix sometimes covering a seed.
Axil — the upper angle that a petiole makes with the stem.
Axillary — in an axil.

B

Berry — a soft fruit with seeds but no stone.
Biennial — of two seasons duration.
Bipinnate — twice pinnate, the primary leaflets are again pinnately divided.

Bisexual—both sexes are functional in the same flower.
Blade—the expanded part of a leaf.
Bract—a reduced leaf, often scale-like, as seen in a flower cluster.
Bulbil—small bulb-like structure usually in the axil of a leaf.

C

Calyx—the outer whorl of leaves of a flower composed of united or separated sepals.
Campanulate—bell shaped.
Capsule—a dry fruit composed of two or more carpels.
Carpel—one of the units of an ovary.
Cladode—flattened branch taking the form and function of a leaf.
Composite—an apparently simple organ existing of two or more structures. (Compositae)
Compound leaf—a leaf composed of two or more leaflets.
Cordate—heart shaped.
Corolla—inner whorl of floral leaves, composed of petals.
Corymb—flat-topped indeterminate inflorescence.
Crenate—with shallow sharp or rounded teeth.
Culm—the stem of grasses.
Cultivar—horticultural variety originating from and growing under cultivation.
Cuneate—wedge shaped, narrowly triangular with the narrow end at the point of attachment.
Cuspidate—with an apical cusp or sharp, rigid point.
Cyathium—a type of inflorescence where several male flowers surround a single female flower.
Cyme—a determinate inflorescence, usually more or less flat-topped, the center flower opening first.
Cymose—cyme-like, borne in cymes.

D

Dehiscent—bursting open of fruit at maturity.
Deltoid—with abroad base and a sharp apex, like a triangle.

Dentate—sharp coarse teeth perpendicular to the margin.

Determinate—an inflorescence in which the terminal flowers open first.

Dioecious—unisexual, male and female flowers on separate plants.

Disc flower—a tubular flower in the central area of the disc of a composite.

Double—a flower that seems to have extra petals beyond the normal number. Often these flowers are sterile, missing stamens and pistil.

Drupe—a stone fruit. A one-seeded indehiscent fruit with the seed enclosed in a hard endocarp which in turn is enclosed in a soft pericarp.

E

Elliptic—oblong, but narrowed at both ends and widest in the middle.

Ectocarp—the outer layer a fruit wall.

Endocarp—the inner layer a fruit wall.

Endosperm—the starch or oil-rich tissue in many seeds.

Entire—Continuous unbroken margin, not toothed or indented.

F

Filament—thread or stem that bears anthers of stamens.

Filiform—threadlike.

Floret—very small flower of a composite; small flowers in the inflorescence of grasses.

Follicle—dry dehiscent one-carpelled fruit, usually with more than one seed.

Four-merous—all parts of a flower (petals, sepals and anthers) occur in multiples of four.

G

Glabrous—without any hairs.

Glaucous—covered with bloom.

Globose—nearly spherical.

H

Habit—general appearance of a plant, such as tree or herb.
Habitat—the kind of place where a plant grows.
Halophyte—a plant tolerant of salt.
Herb—a plant without woody stems.
Hirsute—Covered with rough coarse hairs.

I

Illegitimate name—a plant name contrary to the rules of the Code of Botanical Nomenclature.
Indehiscent—a fruit that does not burst open at maturity.
Indeterminate—an inflorescence such as a raceme where the growth of the main stem is not stopped by the opening of the first flower.
Inequilateral—the two sides of a leaf are unequal in shape and size.
Inflorescence—the flowering part of a plant.
Internode—part of the stem between nodes.
Irregular flower—a zygomorphic or asymmetrical flower.

K

Keeled—ridged.

L

Lanceolate—lance shaped, several times longer than broad, widest below the middle, tapering towards apex with convex sides.
Latex—milky sap.
Leaflet—one of the ultimate parts of a compound leaf.
Liana—a vigorous, woody vine.
Linear—long, narrow, with parallel sides, as leaves of grasses.

M

Margin—edge of a leaf.
Midrib—the central nerve of a leaf.
Monoecious—male and female flowers, staminate and pistillate flowers, on the same plant.
Mucronate—terminated by an abrupt sharp tip.

N

Node—the place on a stem where one or more leaves are attached.

Nut—an indehiscent, one-celled and one-seeded, hard fruit; a drupe with thin, fleshy part and a large stone.

O

Ob-—prefix meaning inverse or reverse.

Oblanceolate—inversely lanceolate with the broadest part above middle.

Orbicular—flat, but circular.

Ovate—outline like an egg, rounded at both ends, widest below the middle.

Ovary—the female reproductive part of the pistil that contains the ovules, which become the seeds.

Ovule—the part of a plant that contains the egg cell and develops into the seed after fertilization.

P

Palmate—with three or more nerves, leaflets, leaves or lobes, all radiating from a common point of attachment.

Panicle—a branching inflorescence, branches usually with racemes or corymbs.

Papilla—small pimple-like protuberances.

Pedicel—the stem of an individual flower.

Peduncle—the stem of a flower cluster, or of a single flower if the inflorescence consists of one flower.

Perennial—of three or more seasons.

Perianth—collective term of the floral envelopes: calyx, corolla, stamens, pistil.

Pericarp—the wall of a ripened ovary, a fruit, sometimes differentiated into an outer layer, **ectocarp**, and an inner layer, **endocarp**.

Petal—unit of the corolla, inner layer of the floral envelope, usually colored.

Petiolate—having a petiole.

Petiole—stem of a leaf.

Phyllary—a bract, especially of the involucrum of the flower head of a composite.

Pilose—covered, but not densely so, with hairs.

Pinna—primary division or primary leaflet of a pinnately compound leaf. If the leaflet is also pinnately divided the individual leaflet become **pinnules.**

Pinnate—constructed like a feather with the parts arranged on both sides of an axis.

Pinnately—divided in a pinnate manner, but not necessarily of a leaf.

Pistil—unit of the gynoecium consisting of ovary, style, and stigma.

Plumose—feather-like.

Pneumatophore—a modified root functioning as a respiratory organ.

Pod—dehiscent dry fruit (beans).

Prop root—an aerial root that comes from the stem, reaches the ground, and helps to support the stem.

Prostrate—lying flat on the ground.

Pseud(o)—false, not true.

Pseudobulb—a thickened, above ground stem in certain orchids.

Pubescent—covered with soft fine hairs, more commonly just hairy.

R

Raceme—an unbranched, elongated, indeterminate inflorescence with pedicelled flowers. The outer or lower flowers open first.

Racemose—flowers in racemes, borne in a raceme or raceme-like inflorescence.

Rachis—axis of an inflorescence or compound leaf.

Rank—vertical row, leaves that are two-ranked are in two vertical rows, one opposite the other. They may be opposite or alternate.

Ray flower—a ligulate flower, with corolla flattened above a very short tube as in composites. In many species the ray flowers are on the margin of a flower head.

Retuse — slightly notched at a usually obtuse apex.

Rhizome — a horizontal stem under the ground that sends up a succession of stems or leaves.

S

Scale — a dry, small, leaf or bract pressed against the stem.

Schizocarp — a dry dehiscent fruit that splits into two halves.

Scorpioid cyme — coiled determinate inflorescence.

Sepal — one of the separate parts of a calyx, usually green.

Serrate — saw toothed with teeth pointing forward to the apex.

Serrulate — minutely serrate.

Sessile — without a stem.

Sheath — a more or less tubular organ surrounding another organ or part, such as, the basal part of grass and palm leaves.

Shrub — woody plant with branches from the ground up, no single trunk.

Simple — leaves that are not divided; inflorescence that is not branched.

Sinuate — with strong wavy indentations.

Spadix — a fleshy flower spike that is surrounded by a leaf or spathe.

Spathe — a large bract or leaf that subtends and often partially surrounds an inflorescence.

Spatulate — spatula shaped, oblong with the base narrowed and the tip rounded.

Spike — an unbranched, elongated, indeterminate inflorescence in which the flowers are sessile or an inflorescence with sessile, composite heads or otherwise condensed flower clusters.

Sporophyll — a spore-bearing leaf.

Stamen — the pollen-bearing organ of a seed plant, part of the androecium.

Standard — the upper petal of a flower.

Stigma — the apical part of pistil that receives pollen for germination.

Stipule—a leaf appendage at the base of the petiole, usually in pairs.
Stolon—a stem that creeps horizontally over the ground and takes root or gives rise to roots.
Style—part of the pistil betweens stigma and ovary.
Suborbicular—almost orbicular or round.
Subshrub—perennial with woody basal stems.
Succulent—fleshy, juicy, and thick.
Suffrutescent—slightly shrubby with only woody basal parts, a sub-shrub.
Symmetrical—capable of being divided into similar halves, also used for radially symmetrical or actinomorph.

T

Tepal— a segment of a perianth, not differentiated in calyx and corolla.
Tomentose—with tomentum, covered with densely matted, short woolly hairs.
Tree—woody plant with a trunk and a more or less distinct and elevated crown.
Tuber—a short thick, usually subterranean, stem baring buds or "eyes" and serving as a storage organ (potato).
Twining—coiling around plants or objects for support.

U

Umbel—an indeterminate, usually flat-topped or convex inflorescence in which the pedicels originate more or less from same point. A compound umbel is where the peduncles support secondary umbels.

W

Whorl—a circle of three or more leaves, flowers or other organs in one node.
Wing—a dry, flat, thin, membranaceous expansion or appendage of an organ as the wing of a seed.

Z

Zygomorphic—bilaterally symmetrical, capable of being divided into two equal halves in only one direction.

Inventory

This list comprises the names of all plants found, collected and preserved on Caye Caulker from 2005–2009. The order of the families and the nomenclature is that of the *Checklist of the Vascular of the Plants of Belize* by Balick, Nee and Atha (2000) for the New York Botanical Garden.

Only the native orchids and bromeliads are included. Some plants could not be collected because of size or location. A photo record of such plants has been used instead. Some plants are included in this list that were accidental or chance observations and may not be permanent on Caye Caulker. These plants are not described in the text. Some plants are only identified to the genus level and are denoted by "sp."

PINOPHYTA
- **ARAUCARIACEAE**
  - Araucaria heterophylla (Salisb.) Franco
- **CUPRESSACEAE**
  - Platycladus orientalis (L.) Franco

CYCADOPHYTA
- **CYCADACEAE**
  - Cycas sp.

MAGNOLIOPHYTA: *Magnoliopsida*
- **ANNONACEAE**
  - Annona glabra L.
  - Annona muricata L.
  - Annona reticulata L.
- **MORACEAE**
  - Artocarpus altilis (Parkinson) Fosberg
  - Ficus crassinervia Desf. ex Willd.
  - Ficus elastica Roxb. ex Hornem
  - Ficus sp. (possibly Ficus benjamina L.)
  - Morus alba L.
- **URTICACEAE**
  - Pilea microphylla (L.) Liebm.
- **CASUARINACEAE**
  - Casuarina equisetifolia L.
- **PHYTOLACCACEAE**
  - Rivina humilis L.

NYCTAGINACEAE
  Boerhavia diffusa L.
  Mirabilis jalapa L.
  Bougainvillea x buttiana Holttum & Standl.
AIZOACEAE
  Sesuvium portulacastrum (L.) L.
CACTACEAE
  Opuntia sp.
CHENOPODIACEAE
  Chenopodium ambrosioides L.
  Salicornia bigelovii Torr.
  Salicornia perennis P. Mill.
  Suaeda linearis (Elliott) Moquin-Tandon
AMARANTHACEAE
  Alternanthera flavescens H.B.K.
  Amaranthus dubius Thell.
  Blutaparon vermiculare (L.) Mears.
PORTULACACEAE
  Portulaca oleracea L.
POLYGONACEAE
  Antigonon leptopus Hook. & Arn.
  Coccoloba uvifera (L.) L.
PLUMBAGINACEAE
  Plumbago sp.
STERCULIACEAE
  Waltheria indica L.
BOMBACACEAE
  Ceiba pentandra (L.) Gaertn.
  Pseudobombax ellipticum (H.B.K.)
  Pseudobombax ellipticoideum A. Robyns
MALVACEAE
  Hibiscus rosa-sinensis L.
  Malvastrum corchorifolium (Desr.) Britton ex Small
  Malvaviscus arboreus Cav.
  Malvaviscus penduliflorus DC.
  Sida acuta Burm. f.
  Sida rhombifolia L.
TURNERACEAE
  Turnera ulmifolia L.
PASSIFLORACEAE
  Passiflora suberosa L.

CARICACEAE
  Carica papaya L.
CUCURBITACEAE
  Momordica charantia L.
BRASSICACEAE
  Cakile lanceolata (Willd.) O.E. Schulz
  Lepidium virginicum L.
BATACEAE
  Batis maritima L.
SAPOTACEAE
  Chrysophyllum cainato L.
  Manilkara zapota (L.) P. Royen
  Pouteria campechiana (H.B.K.) Baehni
  Sideroxylon americanum (Mill.) T.D. Penn.
CRASSULACEAE
  Kalanchoe blossfeldiana Poelln.
  Kalanchoe pinnata (Lam.) Pers.
ROSACEAE
  Rosa sp.
CHRYSOBALANACEAE
  Chrysobalanus icaco L.
SURIANACEAE
  Suriana maritima L.
FABACEAE: Mimosoideae
  Mimosa pudica L.
  Mimosa tarda Barneby
  Pithecellobium keyense Britton
FABACEAE: Caesalpinioideae
  Bauhinia variegata L.
  Bauhinia sp.
  Caesalpinia pulcherrima (L.) Sw.
  Cassia fistula L.
  Chamaecrista nictitans (L.) Moench. var. jaliscensis
    (Greenm.) H.S. Irwin & Barneby
  Delonix regia (Boyer ex. Hook.) Raf.
  Parkinsonia sp.
  Senna alata (L.) Roxb.
  Tamarindus indica L.
FABACEAE: Papillionoideae
  Canavalia rosea (Sw.) DC.
  Crotalaria retusa L.
  Crotalaria verrucosa L.

Desmodium incanum DC.
Desmodium scorpiurus (Sw.) Desv.
Desmodium tortuosum (Sw.) DC.
Erythrina variegata L.
Gliricidia sepium (Jacq.) Kunth ex Walp.
Sesbania grandiflora (L.) Pers.
Sophora tomentosa L.
Vigna luteola (Jacq.) Benth.

LYTHRACEAE
Lagerstroemia indica L.

MYRTACEAE
Eugenia uniflora L.
Psidium guajava L.
Syzygium cumini (L.) Skeels

ONAGRACEAE
Ludwigia octovalvis (Jacq.) P.H. Raven

COMBRETACEAE
Conocarpus erecta L.
Laguncularia racemosa (L.) C.F. Gaertn.
Terminalia catappa L.

RHIZOPHORACEAE
Rhizophora mangle L.

LORANTHACEAE
Struthanthus cassythoides Millsp. ex Standl.

CELASTRACEAE
Crossopetalum rhacoma Grantz.

EUPHORBIACEAE
Acalypha amentacea Roxb. subsp. wilkesiana (Müll. Arg.) Fosberg
Acalypha hispida Burm. f.
Breynia disticha J.R. Forst. & G. Forst.
Chamaesyce blodgettii (Engelm. ex Hirchc.) Small.
Chamaesyce cozumelensis Millsp.
Chamaesyce hypericifolia (L.) Millsp.
Chamaesyce hyssopifolia (L.) Small.
Chamaesyce mesembrianthemifolia (Jacq.) Dugand
Cnidoscolus chayamansa McVaugh
Codiaeum variegatum (L.) A. Juss.
Euphorbia heterophylla L.
Euphorbia tirucalli L.
Jatropha integerrima Jacq.
Manihot esculenta Crantz

Pedilanthus tithymaloides (L.) Poit.
Phyllanthus acidus (L.) Skeels
Phyllanthus amarus Schumach.
Phyllanthus niruri L.
Ricinus communis L.
RHAMNACEAE
Ziziphus mauritiana Lam.
MALPIGHIACEAE
Byrsonima crassifolia (L.) H.B.K.
Galphimia glauca Cav.
SAPINDACEAE
Melicoccus bijugatus Jacq.
BURSERACEAE
Bursera simaruba (L.) Sarg.
ANACARDIACEAE
Mangifera indica L.
Metopium brownei (Jacq.) Urb.
Spondias dulcis Sol. ex Parkinson
Spondias purpurea L.
MELIACEAE
Azadirachta indica A. Juss.
RUTACEAE
Citrus aurantifolia (Christm.) Swingle
Citrus limon (L.) Burm. f.
Ruta sp.
Triphasia trifolia (Burm. f.) P. Wilson
ARALIACEAE
Polyscias fruticosa (L.) Harms
Polyscias tricochleata (Miq.) Fosb.
Schefflera sp.
LOGANIACEAE
Spigelia anthelmia L.
GENTIANACEAE
Eustoma exaltatum (L.) Salisb.
APOCYNACEAE
Allamanda blanchetii A. DC.
Allamanda cathartica L.
Catharanthus roseus (L.) G. Don.
Echites umbellata Jacq.
Nerium oleander L.
Plumeria obtusa L.
Plumeria rubra L.

Rauvolfia tetraphylla L.
Rhabdadenia biflora (Jacq.) Müll. Arg.
Thevetia peruviana (Pers.) K. Schum.
ASCLEPIADACEAE
Asclepias curassavica L.
Cryptostegia grandiflora R. Br.
SOLANACEAE
Brugmansia suaveolens (Humb. & Bonpl. ex Willd) Bercht. & J. Presl.
Datura metel L.
Nicotiana tabacum L.
Solanum americanum Mill.
Solanum donianum Walp.
CONVOLVULACEAE
Ipomoea indica (Burm. f.) Merr.
Ipomoea batatas (L.) Lam
Ipomoea carnea Jacq.
Ipomoea pes-caprae (L.) R. Br.
Ipomoea triloba L.
Merremia dissecta (Jacq.) H. Hallier f.
BORAGINACEAE
Cordia dodecandra A. DC.
Cordia sebestena L.
Heliotropium curassavicum L.
Tournefortia gnaphalodes (L.) Roem. & Schult.
VERBENACEAE
Avicennia germinans (L.) L.
Clerodendrum speciosissimum C. Morren
Clerodendrum thomsoniae Balf.
Duranta repens L.
Lantana camara L.
Lantana involucrata L.
Lippia alba (Mill.) N.E. Br.
Lippia graveolens H.B.K.
Petraea volubilis L.
Phyla nodiflora (L.) Greene
Priva lappulacea (L.) Pers.
Stachytarpheta cayannensis (Rich.) Vahl
LAMIACEAE
Hyptis pectinata (L.) Poit.
OLEACEAE
Jasminum grandiflorum L.
Jasminum multiflorum (Burm. f.) Andrews.

SCROPHULARIACEAE
Bacopa monnieri (L.) Wettst.
Capraria biflora L.
Russelia equisetiformis Schltdl. & Cham.
Russelia sarmentosa Jacq.
Stemodia maritima L.

ACANTHACEAE
Megaskepasma erythrochlamys Linday
Pseuderanthemum carruthersii (Seem.) Guillaumin
Ruellia brittoniana Leonard.

BIGNONIACEAE
Mansoa hymenaea (DC.) A.H. Gentry
Spathodea campanulata P. Beauv.
Tabebuia rosea (Bertol.) DC.
Tecoma stans (L.) H.B.K.

GOODENIACEAE
Scaevola plumieri (L.) Vahl

RUBIACEAE
Chiococca alba (L.) Hitchc.
Erithalis fruticosa L.
Ernodea littoralis Sw.
Hamelia patens Jacq.
Ixora coccinea L.
Morinda citrifolia L.
Mussuenda phillipica A. Rich
Oldenlandia corymbosa L.
Spermacoce verticillata L.
Strumpfia maritima Jacq.

CAPRIFOLIACEAE
Sambucus mexicana C. Presl ex DC.

ASTERACEAE
Ageratum littorale A. Gray
Bidens pilosa L.
Borrichia arborescens (L.) DC.
Conyza canadensis (L.) Cronquist
Cyanthillium cinereum (L.) H. Rob.
Eclipta prostrata (L.) L.
Melanthera nivea (L.) Small
Pluchea carolinensis (Jacq.) G. Don
Pluchea odorata (L.) Cass.
Porophyllum punctatum (Mill.) S.F. Blake
Pseudogynoxys chenopodioides (H.B.K.) Cabrera
Sphagneticola trilobata (L.) Pruski
Tridax procumbens L.

276 Plants of Caye Caulker

MAGNOLIOPHYTA: Lilliopsida
  ASPARAGACEAE
    Asparagus plumosus Baker.
  DRACAENACEAE
    Dracaena americana Donn. Sm.
    Dracaena reflexa Lam.
    Dracaena fragrans (L.) Ker-Gawl. cv Massangeana
    Sansevieria hyacinthoides (L.) Druce
  ASTELIACEAE
    Cordyline fruticosa (L.) A. Chev.
  AGAVACEAE
    Agave sp.
  ASPHODELACEAE
    Aloe vera (L.) Burm. f.
  AMARYLLIDACEAE
    Crinum asiaticum L.
    Crinum sp.
    Hymenocallis littoralis (Jacq.) Salisb.
    Zephyranthes lindleyana Herb.
    Zephyranthes citrina Baker.
  IRIDACEAE
    Cipura campanulata Ravenna
  ORCHIDACEAE
    Myrmecophila sp.
  ARACEAE
    Alocasia macrorrhizos (L.) G. Don
  HYDROCHARITACEAE
    Thalassia testudinum K.D. Koenig
  CYMODOCEACEAE
    Syringodium filiforme Kütz.
  BROMELIACEAE
    Tillandsia streptophylla Scheidw. ex E. Morren
  TYPHACEAE
    Typha domingensis Pers.
  MUSACEAE
    Musa x paradisiaca L.
  HELICONIACEAE
    Heliconia hirsuta L. f.
  STRELITZIACEAE
    Ravenala madagascariensis Sonn.

CANNACEAE
Canna x generalis L.H. Bailey
COMMELINACEAE
Tradescantia spathacea Sw.
CYPERACEAE
Cladium jajmaiscense Crantz
Cyperus ligularis L.
Cyperus polystachyos Rottb.
Eleocharis sp.
Fimbristylis cymosa R. Br.
Fimbristylis spadicea (L.) Vahl
POACEAE
Andropogon glomeratus (Walter) Britton, Sterns & Poggenb.
Cenchrus echinatus L.
Cenchrus incertus M.A. Curtis.
Cymbopogon citratus (DC.) Stapf.
Cynodon dactylon (L.) Pers.
Dactyloctenium aegyptium (L.) Willd.
Distichlis spicata (L.) Greene
Eleusine indica (L.) Gaertn.
Eragrostis amabilis (L.) Wight & Arn. ex Nees
Eragrostis elliottii S. Watson
Eragrostis prolifera (Sw.) Steud.
Eustachys petraea (Sw.) Desv.
Panicum altum Hitchc. & Chase
Paspalum blodgettii Chapm.
Paspalum langei (E. Fourn.) Nash
Paspalum virgatum L.
Saccharum officinarum L.
Spartina spartinae (Trin.) Merr. ex Hitchc.
Sporobolis virginicus (L.) Kunth
Zoysia matrella (L.) Merr. var. matrella
ARECACEAE
Caryota sp.
Chamaedorea seifrizii Burret
Cocos nucifera L.
Phoenix dactylifera L.
Pritchardia pacifica Seem. ex H. Wendl.
Thrinax radiata Lodd. ex. Schult. & Schult. f.
Veitchia merrillii (Becc.) H.E. Moore
PANDANACEAE
Pandanus sp.

References

(Websites accessed March 21, 2009)

Bailey, L.H. and E.Z. Bailey, 1976. *Hortus Third: A Concise Dictionary of Plants Cultivated in the United States and Canada.* New York: MacMillan Publ. Co.

Balick, M.J., M.H. Nee and D.E. Atha, 2000. Checklist of the Vascular Plants of Belize. *Memoirs of the New York Botanical Gardens*, Vol. 85.

Biodiversity & Environmental Resource Data System (BERDS). <biodiversity.bz>

Carrington, S., 1998. *Wild Plants of the Eastern Caribbean.* New York: MacMillan Education Ltd.

Craig, Alan, 1966. *Geography of Fishing in British Honduras and Adjacent Coastal Areas*, M.S. Thesis, Louisiana State University, Baton Rouge LA.

Desert-Tropicals. <desert-tropicals.com>

Duke, J. A, 1983. *Handbook of Energy Crops.* <hort.purdue.edu/newcrop/duke_energy/dukeindex.html>

Earle, C.J. ed. *The Gymnosperm Database.* <conifers.org>

Encyclopedia of Plants and Flowers. <botany.com>

Ethnobotany and Floristics of Belize. <nybg.org/bsci/belize>

Fernald, M.L., 1950. *Gray's Manual of Botany.* 8th Ed. American Book Company. New York.

Flora of China. <efloras.org>

Flora of North America Association, 2007. *Flora of North America.* <efloras.org>

Flora online. <plantnet.rbgsyd.new.gov.au/cgi-bin/NSWfl>

Floridata. <floridata.com>

Heywood, V.H., ed., 1978. *Flowering Plants of the World.* New York: Mayflower Books, Inc.

Hitchcock. A.S., 1950. *Manual of the Grasses of the United States*, 2nd Ed. Revised by A.Chase Dover Publications, Inc.

Honeychurch, P.N., 1980. *Caribbean Wild Plants and their Uses.* New York: MacMillan Education Ltd.

Marin, Isela, 2006. Interview, March 6, 2006.

Marin, Isela, 2007. Interview, February 22, 2007.

Marin, Isela, 2008. Interview, June 20, 2008.

Morton, J.F., 1987. Fruits of Warm Climates. Winterville NC: Creative Resource Systems, Inc.

Nellis, David W., 1994. *Seashore Plants of South Florida and the Caribbean*, Sarasota FL: Pineapple Press, Inc.

Novelo, Ricardo, 2008. Interview, July 13, 2008.

Ocean Oasis Field Guide. <oceanoasis.org/fieldguide>

Pacific Island Ecosystems at Risk (PIER). <hear.org>

Perez, Aurora, 2006. Interview February 26, 2006.

Perez, Aurora, 2008. Interviews June 20 and November 7, 2008.

Rosatti, T., ed., 2008. *The Jepson Manual: Vascular Plants of California*. <ucjeps.berkeley.edu/jepsonmanual/review/>

Scurlock, J. Paul, 1987. *Native Trees and Shrubs of the Florida Keys*, Lower Sugar Loaf Key FL: Laurel and Herbert, Inc.

Standley, P. and J.A.Steyermark. 1946. *Flora of Guatemala*. Chicago IL: Chicago Natural History Museum.

Stoddart, D. R., et. al., 1982. *Cayes of the Belize Barrier Reef and Lagoon System*, Washington, D.C.: Smithsonian Institution.

Vega, Lydia, 2006. Interview, August 13, 2006.

Vega, Lydia, 2008. Interview November 4, 2008.

Watson, L. and Dalwitz, M.J., 2006. The families of flowering plants: descriptions, illustrations, identification, and information retrieval. <delta-intkey.com>

Wikipedia, the Free Encyclopedia. <en.wikipedia.org>

Wildflowers of the Southeastern United States. <2bnTheWild.com>

Young, Peter, Sr., 2008. Interview October 30, 2008.

Zisman, Simon, 1992. *Mangroves in Belize: Forest Planning and Management Project for the Belize Forest Department, Belmopan, Belize*. Belize: Belize Government Printery.

280   *Plants of Caye Caulker*

Index

Scientific names are in italics. Families are in all capital letters.
Pages with the plant description and photo are in **bold**.

A

a'bl po'k **200**
Acalypha amentacea **114**, 272
Acalypha hispida **137**, 272
ACANTHACEAE 98, 275
aceite' **170**
acerola **175**
Aegyptian grass **241**
African tulip tree **199**
AGAVACEAE **159–160**, 276
agave **159–160**
Agave sp. **159–160**, 276
Agave angustifolia **159–160**
Ageratum littorale **60**, 275
Aguilar, Louis A. viii, 28, 260
ah cha'al che' **92**
air plant **27**
AIZOACEAE 39, 270
a'k top' **150**
alder, yellow **110**
allamanda 217, **218**
allamanda, purple **150**
Allamanda blanchetii **150**, 216, 273
Allamanda cathartica **217**, **218**, 273
almond tree, Indian **205**
Alocasia macrorrhizos **161**, 276
aloe **162**
Aloe vera **162**, 276
Alternanthera flavescens **115**, 270
ambarella **200**
amapola **194**
AMARANTHACEAE 32, 115, 155, 270
Amaranthus dubius **155**, 270
AMARYLLIDACEAE 77, 120, 149, 276
anacahuita **48**
ANACARDIACEAE 54, 188, 200, 201, 273
Andropogon glomeratus **237**, 277
angel's bell **117**
angel's trumpet **117**
ankitz **136**
Annona glabra **172**, 269

Annona muricata **269**
Annona reticulata **269**
ANNONACEAE 172, 269
Antigonon leptopus **219**, 270
apasote **157**
apple, gub **172**
apple, pond **172**
apple, star **179**
APOCYNACEAE 37, 74, 96, 136, 145, 150, 151, 193, 218, 273
ARACEAE 161, 276
ARALIACEAE 123, 197, 273
Araucaria heterophylla **173**, 269
ARAUCARIACEAE 173, 269
ARECACEAE 58, 210–215, 277
Armstrong, Ellen viii
artillery plant **90**
Artocarpus altilis **174**, 269
ASCLEPIADACEAE 61, 152, 274
Asclepias curassavica **61**, 274
ASPARAGACEAE 220, 276
asparagus fern **220**
Asparagus plumosus **220**, 276
ASPHODELACEAE 162, 276
ASTELIACEAE 163, 276
ASTERACEAE 33, 35, 60, 62, 67, 70, 83, 91-3, 104, 109, 232, 275
Atha, Daniel viii, 20, 28
Australian pine **178**
Avicennia germinans **22**, 274

B

Bacopa monnieri **30**, 275
ba'eneno che' **61**
ba'ega che' **109**
bamboo palm **210**
banana **122**
bark 7
BATACEAE 31, 271
Batis maritima **31**, 271
Bauhinia sp. 271
Bauhinia variegata L. 271
bay cedar **41**

beach bean **221**
beach morning glory **78**
beach pea **69**
bean, sea **221**
bearded fig **52**
beggarweed **73**
bell, angel's **117**
bell, little **224**
Bermuda grass **240**
Beveridge, Dorothy viii, 28, 260, 289
bicaria **151**
Bidens pilosa **62**, 275
BIGNONIACEAE 135, 199, 203, 227, 275
bird of paradise heliconia **132**
bitter cucumber **229**
bitter melon **229**
biota **169**
black mangrove 21, **22**, 29
black poisonwood **54**
black torch **50**
blackberry **202**
bleeding heart vine **222**
Blodgett's crown grass **248**
bloodberry **97**
blossom berry **50**
blue love grass **244–245**
blue vervain **106**
bluebell **75**
bluebell, Mexican **98**
Blutaparon vermiculare **32,** 270
Boerhavia coccinea **63**
Boerhavia diffusa **63**, 270
BOMBACACEAE 194, 270
BORAGINACEAE 14, 36, 48, 108, 274
Borrichia arborescens **33**, 275
botan **214**
botan blanco **103**
Bougainvillea x buttiana **116**, 270
bougainvillea 113, **116**
bra'ha, ix **134**
bramhi **30**
BRASSICACEAE 34, 79, 271
breadfruit **174**
Breynia disticha **156**, 272
broom grass **237**
broom weed **99**
BROMELIACEAE 27, 276

Brugmansia suaveolens **117**, 274
Bryophyllum pinnatum **158**
bu'nend che' **61**
burr grass **238**
burr, sand **238**
burr-burr **238**
Bursera simaruba **44**, 273
BURSERACEAE 44, 273
buttercup **218**
button sedge **258**
button weed **103**
buttonwood 21, **23**
buu'l che' **55**
Byrsonima crassifolia **175**, 273

C

CCBTIA ii-v, 1, 291
CACTACEAE 167, 270
cactus, pencil **166**
Caesalpinia pulcherrima **130,** 271
cai'mita, ix **179**
caimito **179**
Cakile lanceolata **34**, 271
calalo, ix **155**
calaloo **155**
campana top' **117**
canan **76**
ca'nan, ix **76**
Canavalia rosea **221**, 271
cancerillo **61**
canistel **56**
canna lily **131**
Canna x generalis **131**, 277
CANNACEAE 131, 277
Capraria biflora **64**, 275
CAPRIFOLIACEAE 125, 275
caramelo **56**
Caribbean sedge **255**
Carica papaya **176**, 271
CARICACEAE 176, 271
cascabel **69**
Cassia fistula **177**, 271
castor bean **170**
casuarina **178**
Casuarina equisetifolia **178**, 269
CASUARINACEAE 178, 269
catalina **152**
cat's tongue **95**
Catharanthus roseus **151**, 273

cattail **29, 42, 137**
Caye Caulker 2–7
  climate 2
  cross-section 5
  development 4
  geography 2
  history 4
  map 3
  reserves 7
Caye Caulker Branch, Belize Tourism Industry Association ii-v, 1, 291
cebolla top' **77, 149**
cebollina **149**
cedar, bay **41**
Ceiba pentandra 270
CELASTRACEAE 49, 272
Cenchrus incertus **238**, 277
Cenchrus echinatus **238**
century plant **159**
cha' ya **118**
cha' yu'uk, ix **199**
cha'al che', ix **91**
chac sic **130**
Chamaecrista nictitans 271
Chamaedorea seifrizii **210**, 277
Chamaesyce blodgettii **65–66**, 272
Chamaesyce cozumelensis **65–66**, 272
Chamaesyce hypericifolia **65–66**, 272
Chamaesyce hyssopifolia **65–66**, 272
Chamaesyce mesembrianthemifolia **65–66**, 272
chanca piedra **89**
chaya **118**
cha'ya **118**
chechem **54**
chenille plant **137**
CHENOPODIACEAE 38, 40, 157, 270
Chenopodium ambrosioides **157**, 270
cherry, Indian **207**
cherr, wild y **49**
chestnut sedge **259**
chi **175**
chi chi be **99**
chik íchtup, ix **148**
chicle **189**
children weed **89**
Chiococca alba **45**, 275
Chinese arborvitae **169**
chit **58**

Christmas flower **37**
Christmas palm **215**
CHRYSOBALANACEAE **46**, 271
Chrysobalanus icaco **46**, 271
Chrysophyllum cainito **179**, 271
chufa **255**
cimaron **53**
cina'an top' **137**
Citrus aurantifolia **180**, 273
Citrus limon 273
Cladium jamaiscense **254**
clavo de oro **110**
claudiosa **64**
Clerodendrum speciosissimum **138**, 274
Clerodendrum thomsoniae **222**, 274
climate 2
cloud berry **60**
Cnidoscolus chayamansa **118**, 272
coat buttons **109**
Coccoloba uvifera **47**, 270
cochineal **167**
coco **161, 211**
coconut palm 14, 113, 209, **211–2**
cocoplum **46**
Cocos nucifera 14, **211–2**, 277
Codiaeum variegatum **119**, 272
colcho **172**
COMBRETACEAE 23, 24, 129, 205, 272
COMMELINACEAE 129, 277
common ruellia **98**
Conocarpus erecta **23**, 272
CONVOLVULACEAE 78, 223, 224, 228, 274
Conyza canadensis **67**, 275
coral plant **148**
coral tree **183**
coral vine **219**
Cordia dodecandra **48**, 274
Cordia sebestena 14, **48,** 274
Cordyline fruticosa **163**, 276
cordgrass **251**
corn plant **164**
cortadera **255**
cough bush **91**
cow pea, wild **111**
cowhorn orchid **25**
crabgrass **243**

craboo **175**
CRASSULACEAE **142**, 158, 271
crepe myrtle **154**
crinoline **114**
Crinum asiaticum **120**, 276
Crinum sp. **120**, 276
Crossandra sp. **274**
Crossopetalum rhacoma **49**, 272
Crotalaria retusa **68**, 271
Crotalaria verrucosa **69**, 272
croton, garden **119**
crowfoot grass **241**
Cryptostegia grandiflora **152**, 274
CUCURBITACEAE **229**, 271
CUPRESSACEAE **169**, 269
cure-for-all **91**
cutting grass **251**, **255**
Cyanthillium cinereum **70**, 275
CYCADACEAE **181**, 269
CYCADOPHYTA 269
Cycas sp. **181**, 269
Cymbopogon citratus **239**, 277
CYMODOCEACEAE **18**, 276
Cynodon dactylon **240**, 277
CYPERACEAE 254–259, 277
Cyperus ligularis **255**, 277
Cyperus papyrus 253
Cyperus polystachyos **256,** 277

𝒟

daisy **109**
daisy, seaside **33**
date palm **213**
Dactylopius coccus **167**
Dactyloctenium aegyptium **241**, 277
Delonix regia **182**, 271
Desmodium incanum **71**, 272
Desmodium scorpiurus **72**, 272
Desmodium tortuosum **73**, 272
development **4**
devil pepper **96**
devil's potato **74**
dew drop **153**
Distichlis spicata **242,** 277
Dominican love grass **244–245**
downy jasmine **226**
Dracaena americana 276
Dracaena fragrans **164**, 276
Dracaena reflexa **165**, 276

Index 283

DRACAENACEAE 126,164, 165, 276
Duranta repens **153**, 274
durmelon **84**

ℰ

Echites umbellata **74**, 273
Eclipta prostrata **35**, 275
eel grass **18**
egg fruit **56**
elder, yellow **135**
elderberry, Mexican **125**
Eleocharis sp. **257**, 277
elephant's ear **161**
Elliot's love grass **244–245**
Eleusine indica **243,** 277
epazote **157**
epiphytes 21, 25, 27
Eragrostis **244–245**
Eragrostis amabilis **244–245**, 277
Eragrostis elliottii **244–245**, 277
Eragrostis prolifera **244–245**, 277
Erithalis fruticosa **50**, 275
Ernodea littoralis **51**, 275
Erythrina variegata **183**, 272
Eugenia uniflora **184**, 272
Euphorbia heterophylla 272
Euphorbia tirucalli **166**, 272
EUPHORBIACEAE 65, 89, 114, 118, 119, 137, 141, 146, 156, 166, 170, 192, 272
Eustachys petraea **246**, 277
Eustoma exaltatum **75**, 273

ℱ

FABACEAE: Caesalpinioideae 15, 102, 130,134, 177, 182, 204, 271
FABACEAE: Mimosoideae 15, 55, 84, 85, 271
FABACEAE: Papillionoideae 15, 68-69, 71-73, 111, 183, 187, 198, 221, 271
false cassava **183**
false daisy **35**
false eranthemum **124**
false mallow **82**
false oleander **136**
false primrose **81**
fan flower **127**
fern, asparagus **220**

Ficus crassinervia **52**, 269
Ficus elastica **185**, 269
Ficus sp. 15, **186**, 269
fig 15, **52**, **186**
Fimbristylis castanea **259**
Fimbristylis cymosa **258**, 277
Fimbristylis spadicea **259**, 277
finger grass **246**
firecracker plant **148**
flamboyant **182**
flaming katy **142**
flamingo beak tree **198**
flat-spiked sedge **256**
flor de mayo **203**
flowers 7,9, 9–13
four o'clock **86**
fox tail **237**
frangipani **193**
fruits 13

G

Galphimia glauca **273**
garden croton **119**
garlic vine **227**
geiger tree **48**
GENTIANACEAE 75, 273
geography 2
glasswort **38**
Gliricidia sepium **187**, 272
glory bower **222**
goat weed **64**
golden creeper **51**
golden shower **177**
GOODENIACEAE 127, 275
goose grass **243**
governor's plum **207**, 216
grape, sea **47**
GRAMINEAE 237–252
grasses 7, 17, 235–6
grey mangrove 21, **23**
grossea **192**
guard plant **233**
guava **195**
guayaba **195**
Guazuma ulmifolia 41
gub apple **172**
guinea **122**
gumbolimbo **44**

H

Hamelia patens **76**, 275
hamans **205**
Heliconia 132
Heliconia hirsuta **132**, 276
HELICONIACEAE 132, 276
heliotrope, seaside **36**
Heliotropium curassavicum **36**, 274
Heliotropium gnaphalodes **108**
herbs 7
Heredia, Lionel (Chocolate) 208
hibiscus **113, 139**
Hibiscus rosa-sinensis **139**, 270
hicaco **46**
higuerilla **170**
history 4
hog plum **200**
hogweed, little **94**
hook-and-eye lily **77**
horseweed **67**
HYDROCHARITACEAE 19, 276
Hymenocallis littoralis **77**, 276
Hyptis pectinata 274

I

Identification of plants 1, 14
ik tulipan, ix **144**
Indian almond tree **205**
Indian cherry **207**
Indian jujube **207**
Indian plum **207**
inflorescence 9, 12–13
ink plant **50**
inkberry **127**
Ipomoea indica **223**, 274
Ipomoea pes-caprae **78**, 274
Ipomoea triloba **224**, 274
ironweed, little **70**
island romero **128**
ix bra'ha **134**
ix cai'mita **179**
ix calalo **155**
ix ca'nan **76**
ix cha' yu'uk **199**
ix cha'al che' **91**
ix chik íchtup **148**
ix ik tulipan **144**
ix mu'tz **84**

ix si'nki'n **130**
ix tai' tai' che' **93**
ix tze'kel ba'ak **86**
ix vervaina **106**
ixike' **126, 159–60**
Ixora coccinea **140**, 275
ixsi'k tze'kel ba'ak **124**
ixora **140**

J

Jacob's coat **114**
jaguey **52**
jambolan **202**
Japanese carpet grass **252**
Japanese poinsettia **146**
jasmine **74, 225, 226**
jasmine, downy **226**
jasmine, star **226**
Jasminum grandiflorum **225**, 274
Jasminum multiflorum **226**, 274
Jasminum officinale **225**
jatropha **141**
Jatropha integerrima **141**, 272
java plum **202**
jocote **201**
Johnston, Allie viii
Joseph, Petrona **234**
joyweed **115**
jugs, yellow **51**
jujube, Indian **207**
jungle geranium **140**

K

Kalanchoe blossfeldiana **142**, 271
Kalanchoe pinnata **158**, 271
kenep **190**

L

laburnum **177**
Lagerstroemia indica **154**, 272
Laguncularia racemosa **24**, 272
LAMIACEAE **274**
lang poul **87**
lantana **61, 133**
Lantana camara **133**, 274
Lantana involucrata **53**, 274
lavender, sea **108**
leaf structure **8–11**
leaves **8–11**

leche' che' **89**
lemon grass **239**
Lepidium virginicum **79**, 271
life everlasting **158**
lily, milk and wine **120**
lily, rain **149**
lily, red **120**
lily, spider **77**
lime **180**
lime berry **206**
lime, Spanish **190**
li'mo'n **180**
limonaria **206**
Lippia alba **80,** 274
Lippia graveolens **121**, 274
Lippia nodiflora **88**
little bell **224**
little hogweed **94**
little ironweed **70**
LOGANIACEAE **105**, 273
LORANTHACEAE **233**, 272
love grass **244–245**
lover's vine **219**
Ludwigia octovalvis **81,** 272
Lumb, Judy viii
LYTHRACEAE **154**, 272

M

madre de cacao **187**
MAGNOLIOPHYTA: Magnoliopsida **269**
MAGNOLIOPHYTA: Lilliopsida **276**
MALPIGHIACEAE **175**, 273
MALVACEAE **82, 99, 139, 143,** 270
Malvastrum corchorifolium **82**, 270
Malvaviscus arboreus **143, 144,** 270
Malvaviscus penduliflorus **143, 144,** 270
mamoncillo **190**
manatee grass 17, **18**
Mangifera indica **188**, 273
mangle **22, 26**
mango **188**
mangrove 21
mangrove rubber vine **37**
mangroves **21–7**
Manihot esculenta 273
Manilkara zapota **189**, 271
Manila grass **252**

Manila palm **215**
Mansoa hymenaea **227**, 275
Marin, Ernesto, Jr. 260
Marin, Isela 28, 260
Marin, Lillianna viii
marshmallow **112**
Massangeana **164**
masapan **174**
mata polo **233**
maxapan **174**
May plum **201**
mayflower **182**
mayflower tree **203**
McRae, Ellen viii
mekl **161**
Melanthera nivea **83**, 275
Melicoccus bijugatus **190**, 273
Merremia dissecta **228**, 274
Metopium brownei **54**, 273
Mexican bluebell **98**
Mexican elderberry **125**
Mexican flame vine **232**
Mexican oregano **121**
Mexican petunia **98**
Mexican weed **157**
milk and wine lily **120**
milkberry **45**
Miller, Mo viii
Mimosa pudica **84**, 271
Mimosa tarda **85**, 271
ming tree **123**
Mirabilis jalapa **86**, 270
mombin, red **201**
mombin, yellow **200**
Momordica charantia **229**, 271
MORACEAE 52, 174, 185, 186, 269
Morinda citrifolia **191**, 275
morning glory **223**
Morus sp. **269**
Moses-in-the-boat **129**
Moses-in-the-cradle **129**
mother-in-law tongue **126**
mul ché **57**
mu'ul che' **57**
Musa x paradisiaca **122**, 276
MUSACEAE 122, 276
Mussuenda phillipica 275
mu'tz, ix **84**

Myrmecophila sp. **25**, 276
MYRTACEAE 176, 184, 195, 201, 272

N

naked lady **166**
names of plants 14–15
nance **175**
napoleona **140**
narcisso **145**
necklace pod **102**
necklace, yellow **102**
needles, Spanish **62**
Nerium oleander **145**, 273
New York Botanical Garden iv, 1, 14
Nicotiana tabacum 274
nightshade **100**
noni tree **191**
nopal **167**
Norfolk Island pine 171, **173**
Novelo, Angelica 216
NYCTAGINACEAE 63, 86, 116, 270

O

Oldenlandia corymbosa **87**, 275
OLEACEAE 225, 274
oleander **113, 145**
oleander, yellow **136**
ONAGRACEAE 81, 272
oole' che' **185**
Opuntia sp. **167**, 270
ORCHIDACEAE 25, 276
ora'ego **121**
ora'ego che' **133**
oregano **80, 121**
oregano, Mexican **121**
oregano, sweet **53**
oregano, wild **133**
ornamental cassava **183**
Otaheite apple **200**
oxeye, seaside **33**

P

pagoda flower **138**
palm lily **163**
palm, Manila **215**
palm, reed **210**
palm, thatch **214**
palms 209–17

Index   287

palmetto, saltwater **58**
PANDANACEAE **168**, 277
Pandanus sp. **168**, 277
pandanus **168**
panicum **247**
Panicum altum **247**, 277
papaya **176**
Parkinsonia sp. **271**
parrot beak tree **198**
Paspalum blodgettii **248**, 277
Paspalum langei **248**, 277
Paspalum virgatum. **249**, 277
Passiflora suberosa **230**, 270
PASSIFLORACEAE **230**, 270
passionflower **230**
paw-paw **176**
pazote' **157**
pea shooter grass **247**
pech ma'am **95**
Pedilanthus tithymaloides **146**, 273
pencil cactus **166**
pepper, devil **96**
pepper grass **79**
Perez, Aurora **216**, 260
periwinkle **151**
Petraea volubilis **231**, 274
petunia, Mexican **98**
Phoenix dactylifera **213**, 277
Phyla nodiflora **88**, 274
Phyllanthus acidus **192**, 273
Phyllanthus amarus **89**, 273
Phyllanthus niruri **89**, 273
PHYTOLACCACEAE **97**, 269
pickle weed **31**
pigeon berry **153**
Pilea microphylla **90**, 269
pincha huevo **251**
pine **173**
pink poui **203**
pinkroot **105**
PINOPHYTA **269**
piss-a-bed **134**
Pithecellobium keyense **55**, 271
plant structure **7**
plantain **122**
platanillo **131**
platano **122**
Platycladus orientalis **169**, 269

pleomele **165**
Pluchea carolinensis **91**, 275
Pluchea odorata **92**, 275
plum, golden **200**
plum, governor's **207**, 216
plum, hog **200**
plum, Indian **207**
plum, java **202**
plum, May **201**
plum, Spanish **200**
PLUMBAGINACEAE **270**
Plumbago sp. **270**
Plumeria obtusa **193**, 273
Plumeria rubra **193** 273
POACEAE **237–252**, 277
poisonwood, black **54**
polly redhead **76**
POLYGONACEAE 47, **219**, 270
Polyscias fruticosa **123**, 273
Polyscias tricochleata **273**
pond apple **172**
Porophyllum punctatum **93**, 275
Portulaca oleracea **94**, 270
PORTULACACEAE **94**, 270
potato, devil's **74**
potato tree **101**
potato vine **224**
Pouteria campechiana **56**, 271
prickly pear **167**
pride of Barbados **130**
pra'uska' **122**
Pritchardia pacifica **214**, 277
Priva lappulacea **95**, 274
Pseuderanthemum carruthersii **124**, 275
Pseudobombax ellipticum **194**, 270
Pseudobombax ellipticoideum **194**, 270
Pseudogynoxys chenopodioides **232**, 275
Psidium guajava **195**, 272
p'ta **195**
purple allamanda **150**
purple sedge **255**
purslane, sea **94**
purslane, seaside **39**, 94
pussy tail **137**
put' **176**

Q

queen's wreath 231

R

rabbit's paw 104
rain lily 149
rattle weed 68
Rauvolfia tetraphylla 96, 274
Ravenala madagascariensis 196, 277
red lily 120
red mangrove 21, 26
red mombin 201
reed palm 210
Requena, Annabella 20
reserves 7
Reyes, Claudia 234
Rhabdadenia biflora 37, 274
RHAMNACEAE 207, 273
Rhizophora mangle 26, 272
RHIZOPHORACEAE 26, 272
rice and beans 222
rick-rack plant 114
Ricinus communis 170, 273
Rietsema, Jacob viii, 28, 208, 216, 234, 260, 289
Rivina humilis 97, 269
roadsides 59–112
rocket, sea 34
roots 7
Rosa sp. 147, 271
ROSACEAE 147, 271
rose 147
rosebay 145
royal palm 215
royal poinciana 182
rubber tree 185
rubber vine 74, 152
RUBIACEAE 45, 50, 51, 76, 87, 103, 128, 140, 191, 275
ruellia, common 98
Ruellia brittoniana 98, 275
Ruellia tweediana 98
Russelia equisetiformis 148, 275
Russelia sarmentosa 148, 275
rusty-seed paspalum 248
Ruta sp. 273
RUTACEAE 180, 206, 273

S

sacate' 239
Saccharum officinarum 250, 277
sage, sea 80
sage, wild 53
Sarcocornia perennis 38
Salicornia bigelovii 38, 270
Salicornia perennis 38, 270
salt grass 242
salt marsh fleabane 92
saltwater palmetto 58
saltweed 32
saltwort 29, 31
Sambucus mexicana 126, 275
sand burr 238
Sansevieria hyacinthoides 126, 276
Santa Maria 91, 92
Santiago 219
SAPINDACEAE 190, 273
sapodilla 189
SAPOTACEAE 56, 57, 179, 189, 271
saw grass 254
Scaevola plumieri 127, 275
schefflera 197
Schefflera sp. 197, 273
Schomburgkia sp. 25
scientific names 14–15
scoggineal 167
scorn d'earth 233
scorpion tick trefoil 72
screw pine 168
SCROPHULARIACEAE 30, 64, 107, 148, 275
sea bean 221
sea grape 47
sea lavender 108
sea purslane 94
sea rocket 34
sea sage 80
seaside daisy 33
seaside heliotrope 36
seaside oxeye 33
seaside purslane 39, 94
seaside twintip 107
sedges 237, 253
seeds 13
seepweed 40
Senna alata 134, 271

sensitive plant **84**
Sesbania grandiflora **198**, 272
Sesuvium portulacastrum **39**, 270
shame lady **84**
Shasta daisy **62**
shaving brush tree **194**
shower of gold **177**
shrubs **7**
Sida acuta **99**, 270
Sida rhombifolia **99**, 270
Sideroxylon americanum **57**, 271
siempreviva **158**
silver bush **102**
silverweed **32**
sink-and-bible **162**
si'nki'n, ix **130**
sis che' **220**
skeleton plant **166**
sleeper **84**
sleeping grass **84**
sleeping hibiscus **143**
sleepy head **84**
Smithsonian **126**
snake plant **126**
snow square stem **83**
snowflake plant **156**
SOLANACEAE **100**, 101, 117, 274
Solanum americanum **100**, 274
Solanum donianum **101**, 274
Sophora tomentosa **102**, 272
sorosi **229**
southern cattail **42**
Spanish lime **190**
Spanish needles **62**
Spanish plum **200**
Spartina spartinae **251**, 277
Spathodea campanulata **199**, 275
Spermacoce verticillata **103**, 275
Sphagneticola trilobata **104**, 276
spider lily **77**
spiderling **63**
spiderwort **129**
Spigelia anthelmia **105**, 273
spike grass **257**
spike rush **257**
spinach, wild **155**
Spondias dulcis **200**, 273
Spondias purpurea **201**, 273
Sporobolus virginicus **277**

squirrel's tail **93**
Stachytarpheta cayannensis **106**, 274
star apple **179**
star jasmine **226**
stem **7**, 8
Stemodia maritima **107**, 275
STERCULIACEAE **112**, 270
stone breaker **89**
STRELITZIACEAE **196**, 277
Strumpfia maritima **128**, 275
Struthanthus cassythoides **233**, 272
Suaeda linearis **40**, 270
sugar cane **250**
sunflower **109**
Suriana maritima **41**, 271
SURIANACEAE **41**, 271
Surinam cherry **184**
sweet oregano **53**
sweet pea, yellow **68**
sweet-scent **92**
switchgrass **249**
Syringodium filiforme **18**, 276
Syzygium cumini **202**, 274

T

Tabebuia rosea **203**, 275
Tahitian gooseberry **192**
tai' tai' che', ix **93**
talquezal **249**
tamarind **204**
Tamarindus indica **204**, 271
tanchi **64**
taro **161**
Tecoma stans **135**, 275
Terminalia catappa **205**, 272
Thalassia testudinum **19**, 276
thatch palm **214**
Thevetia peruviana **136**, 274
Thrinax radiata **58**, 277
Thuja orientalis **169**
ti plant **163**
tick trefoil **71, 73**
Tillandsia streptophylla **27**, 276
t'o **250**
top' kimen **163**
torch, black **50**
Tournefortia gnaphalodes **108**, 274
Tradescantia spathacea **129**, 277

traveler's palm **196**
trees **7, 171**
tree datura **117**
tree of life **158**
tridax daisy **109**
Tridax procumbens **109**, 276
Triphasia trifolia **206**, 273
tropical milkweed **61**
trumpet, angel's **117**
trumpet bush **135**
tulipan **139**
Turk's cap **143, 144**
Turnera ulmifolia **110**, 270
TURNERACEAE **110**, 270
turtle grass 17, **19**
twelve o'clock **228**
twelve o'clock prickle **84**
twintip, seaside 107
Typha domingensis **42**, 276
TYPHACEAE 42, 276
tzac ash **166**
tza'k chu' **146**
tze'kel ba'ak, ix **86**

U

u' moch atu'rich **104**
u top' atzunoon **132**
umbrella tree **197**
URTICACEAE **90**, 269

V

Vega, Lydia 260
Vega, Tony, Sr. viii
Veitchia merrillii **215**, 277
velvet leaf **112**
verbena **88, 106**
VERBENACEAE 22, 53, 80, 88, 95, 106, 121, 133, 138, 153, 222, 231, 274
verdolaga **39, 94**
Vernonia cineria 70
vervain **133**
vervain, blue **106**
vervaina, ix **106**
Vigna luteola **111**, 272

W

Waltheria indica **112**, 270
water hyssop **30**

waya **190**
wayfarer's palm **196**
Wedelia trilobata **104**
well zu'uk **251**
wetlands 29
white lily **77**
white mangrove 21, **24**
whitty whitty **192**
whip plant **106**
white weed **60**
wild cherry **49**
wild cow pea **111**
wild oregano **133**
wild sage **53**
wild spinach **155**
willow **136**
willow primrose **81**
worm bush **105**
worm weed 89

X

xate **210**
xo-coi **55**
xa a'n che' **58**

Y

yax xalac che **146**
yellow alder **110**
yellow elder **135**
yellow jugs **51**
yellow mombin **200**
yellow necklace **102**
yellow oleander **136**
yellow sweet pea **68**
yellow zericote **48**
yerba che' **80**
yerba mora **100**
Young, Alex viii

Z

Zephyranthes citrina **149**, 276
Zephyranthes lindleyana **149**, 276
zericote 14, **48**
zericote, yellow **48**
Ziziphus mauritiana **207**, 216, 273
zoysia grass **252**
Zoysia matrella **252**, 277
zu'uk **237**

Caye Caulker Branch
Belize Tourism Industry Association

The Belize Tourism Industry Association (BTIA) is a non-governmental membership, advocacy organization promoting and maintaining sustainable tourism while safeguarding the natural and cultural integrity of Belize for the benefit of all (BTIA Mission Statement <btia.org>).

The Caye Caulker Branch of BTIA (CCBTIA) was established in 1985, soon after the formation of the national BTIA. CCBTIA has supported the development of sustainable tourism on Caye Caulker through planning and tourism policy development, lobbying for infrastructure improvements and increased enforcement of security, and marketing through trade shows and the Internet <gocayecaulker.com>. CCBTIA has provided numerous training programmes for the public in hospitality, capacity-building, and conservation. The CCBTIA Mini-Reserve is the basis for an extensive environmental education programme. For preservation of the Caye's heritage, both historical and ecological, CCBTIA has produced several publications, including this book.

Authors

Jacob Rietsema received his Ph.D. in Botany at the University of Utrecht, the Netherlands. After moving to the United States, he did research at the Genetics Experiment Station at Smith College, after which he moved to private industry continuing work in botany and agricultural projects as well as in management. After his retirement he became involved with land use, environmental law and town planning.

Dorothy Straughn Beveridge received her B.A. in Education from the University of North Carolina at Chapel Hill and taught for 10 years. For the last 25 years, she has been a tour guide and environmental educator in Belize. She is on the Board of Directors of the CCBTIA and a member of the Belize Audubon Society.

www.ingramcontent.com/pod-product-compliance
Lightning Source LLC
Chambersburg PA
CBHW042054290426
44111CB00001B/6